THOMAS HUNT MORGAN

Thomas Hunt Morgan

Morgan

Pioneer of Genetics

IAN SHINE and
SYLVIA WROBEL

Foreword by George W. Beadle

THE UNIVERSITY PRESS OF KENTUCKY

Research for The Kentucky Bicentennial Bookshelf is assisted by
a grant from the National Endowment for the Humanities. Views
expressed in the Bookshelf do not necessarily represent those of the
Endowment.

The University Press of Kentucky
Scholarly publisher for the Commonwealth,
serving Bellarmine University, Berea College, Centre
College of Kentucky, Eastern Kentucky University,
The Filson Historical Society, Georgetown College,
Kentucky Historical Society, Kentucky State University,
Morehead State University, Murray State University,
Northern Kentucky University, Transylvania University,
University of Kentucky, University of Louisville,
and Western Kentucky University.
All rights reserved.

Editorial and Sales Offices: The University Press of Kentucky
663 South Limestone Street, Lexington, Kentucky 40508-4008
www.kentuckypress.com

Cataloging-in-Publication Data is available from
the Library of Congress.

ISBN 978-0-8131-9337-3 (pbk: acid-free paper)

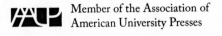

DEDICATED TO
A. S. and M. S. and J. B. and E. B.
whose genes made this book possible

That the fundamental aspects of heredity should have turned out to be so extraordinarily simple, supports us in the hope that nature may, after all, be entirely approachable. Her much-advertised inscrutability has once more been found to be an illusion due to our ignorance. This is encouraging, for if the world in which we live were as complicated as some of our friends would have us believe we might well despair that biology could ever become an exact science.

Thomas Hunt Morgan

Contents

Foreword

SHINE AND WROBEL's masterly work will surely long stand as the definitive biography of Thomas Hunt Morgan the man, citizen, scientist, teacher, collaborator, administrator, husband, parent, and friend. But it is far more than that, for it also chronicles the early history of this century's most significant research achievement in the biological sciences. Morgan, during the more than five decades of his academic career, not only himself contributed notably to that achievement but also provided opportunity for others to further the work.

My own association with him began when he was in his mid-sixties—retirement age for most people, but for Morgan only the beginning of a new chapter. By then he had become a towering figure in American science; yet in many ways he was like Rollins Adams Emerson, with whom I had just taken my Ph.D. degree at Cornell. Morgan's group had made *Drosophila* the genetically best known of animals; Emerson and his associates were doing the same for the plant *Zea mays*, Indian corn. Both men were modest, unpretentious, enthusiastic, creative scholars, especially good at inspiring students and colleagues to extend science through research.

I came to the California Institute of Technology in 1931 as a National Research Council Fellow and stayed, as an instructor, until 1935. Thus I participated in the early years of Morgan's administrative experiment: the creation of an uncompartmentalized division of biological sciences which encouraged the sharing of ideas with biochemists as well as with biologists whose specialty differed from one's own. This kind of organization was almost unique in the country at that time.

After a period spent at other institutions, I returned to Caltech in 1946 (a year after Morgan's death) as chairman of the division he had established two decades earlier. By then, the soundness of his organizational concept had already been demonstrated; all I had to do was maintain it. I left Caltech in 1961 but later served on its board of trustees—so I have been either actively engaged in or observing Morgan-style biology for over forty years.

In the course of Morgan's professional lifetime, the science of genetics was born and developed to the stage where its future depended on interdisciplinary collaboration. The recognition of DNA as the basis of heredity in all higher organisms and the elucidation of its molecular structure and its mechanism of replication initiated a veritable revolution in biological understanding. Today, no self-respecting biologist, biochemist, medical scientist, or even intelligent layman can fail to appreciate the magnitude and significance of that revolution.

Morgan provided it with both personal and institutional impetus. Seven Nobel Prizes have been awarded for biological investigators who at one time or another worked in the division he established. Eleven members of the biology faculty are members of the National Academy of Sciences. To these should be added two members of the Division of Chemistry investigating chemical aspects of biology.

With Robert L. Sinsheimer as the present chairman of the Division of Biology, the area of basic behavioral sciences has been significantly strengthened and expanded. And in collaboration with the Institute's government-supported Jet Propulsion Laboratory, Norman H. Horowitz of the Division of Biology and others are investigating the possibilities of present or past life on the planets of the solar system, at this writing especially on Mars.

In the broader area of world science, the revolution in biology continues to accelerate. As one example, I have just taken a census of the scientific papers in the June 1976 issue of the prestigious *Proceedings* of our National Academy of Sciences. Of the 85 scientific papers in that issue, 76 pertain to basic or medical biology, and of these, 40 are explicitly concerned with genetics.

The scientific legacy of Thomas Hunt Morgan is great indeed, and we have not yet assessed its full magnitude and significance.

George W. Beadle

Preface

ONE HUNDRED YEARS AGO, genetics did not exist. Not only were genes and chromosomes unknown, but the exact role of the sperm and egg in fertilization was not understood. Mankind had not come far from believing that crocodiles arose spontaneously out of the mud of the Nile, worms grew from horsehair, and bacteria sprang newborn out of dirt. Even Charles Darwin, who did so much to disprove the spontaneous creation of species, did not understand how inheritance worked.

Modern genetics began in 1900. It gradually became a science, and with increasing assurance began uncovering nature's mysteries. The mechanical basis of inheritance was laid bare, the way genes duplicated themselves and made proteins was discovered, and eventually the exact structure of the gene itself was revealed. The brotherhood of mankind and indeed the unity of all living things was established.

The exponential rise in Nobel Prizes in Medicine awarded for discoveries in genetics signals that science's new place as the cornerstone of all future biology and the essential language of the educated man, as Latin was a hundred years ago. At present, however, for all its intellectual beauty, the practical benefits of genetics have hardly begun, even though high-yielding disease-resistant crops are already commonplace and rhesus incompatibility, only recently the major cause of stillbirth and brain damage in infants, has largely disappeared.

The first Nobel laureate in genetics, Thomas Hunt Morgan, was born and reared in a well-known Kentucky family. Local schools, including one later to become the University of Kentucky, gave him his basic education. But despite these ties to Kentucky, and despite his being the only Kentuckian to have won the Nobel Prize, Thomas Hunt Morgan remains to Kentuckians one of the least-known of their famous sons.

Morgan seldom returned to the state once he left it. At first he was involved with his work and glad enough to let his family come to him, rather than the other, more time-consuming way around. Later, after most of his family were dead and he himself was famous, the invitations home were usually honorary ones, and he was shy of such occasions. If Kentuckians were hurt that he refused to come to their planned celebration of his seventieth birthday in 1936, they need only recall that he had missed the festive Nobel laureate celebration in Stockholm three years earlier giving the same reason: that he might avoid the speeches in his honor and instead stay with the work at hand.

Morgan was a private man who did not share his life and feelings. But Kentuckians face another barrier. To reach Thomas Hunt Morgan, the scientist, one must somehow get past his uncle, John Hunt Morgan, the Civil War raider, the "Thunderbolt of the Confederacy." The local preoccupation with the uncle was evident even in 1975 at the dedication of the University of Kentucky's new biology building, named, naturally, after Kentucky's most famous biologist; one television reporter happily announced that the building was named in honor of Lexington's famous John Hunt Morgan.

From Kentucky Morgan derived a love of nature, and all his life he studied living things. He was basically a zoologist, as he always tried to make clear, and his abiding interest was to explain the miracle of the ovum's evolution into an adult. Since he did not believe in miracles, he viewed the ovum as a wound-up machine whose bits he intended to take apart and understand. In so doing, he won a Nobel Prize.

In this small book we have presented a neglected man—neglected not only by his fellow Kentuckians but to some extent by his fellow scientists; his spectacular achievements in genetics blinded even specialists to the importance of his embryological work. Having read his books and papers, and especially having talked to his remaining friends, colleagues, and acquaintances and to those who have continued in the science over whose birth Morgan presided, we are convinced that Morgan truly was a great man.

And that's Thomas Hunt.

THE FAMILY TREE OF T

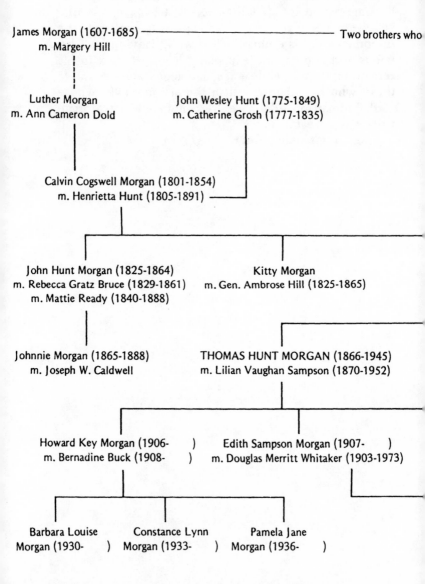

James Morgan (1607-1685) ——————————— Two brothers who
m. Margery Hill

Luther Morgan
m. Ann Cameron Dold

John Wesley Hunt (1775-1849)
m. Catherine Grosh (1777-1835)

Calvin Cogswell Morgan (1801-1854)
m. Henrietta Hunt (1805-1891) ————

John Hunt Morgan (1825-1864)
m. Rebecca Gratz Bruce (1829-1861)
m. Mattie Ready (1840-1888)

Kitty Morgan
m. Gen. Ambrose Hill (1825-1865)

Johnnie Morgan (1865-1888)
m. Joseph W. Caldwell

THOMAS HUNT MORGAN (1866-1945)
m. Lilian Vaughan Sampson (1870-1952)

Howard Key Morgan (1906-   )
m. Bernadine Buck (1908-   )

Edith Sampson Morgan (1907-   )
m. Douglas Merritt Whitaker (1903-1973)

Barbara Louise
Morgan (1930-   )

Constance Lynn
Morgan (1933-   )

Pamela Jane
Morgan (1936-   )

*Abbreviated version

MAS HUNT MORGAN*

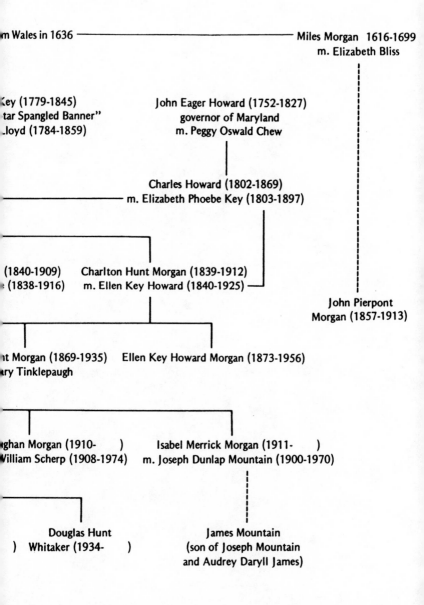

m Wales in 1636 ——————————————————— Miles Morgan 1616-1699
m. Elizabeth Bliss

Key (1779-1845)
tar Spangled Banner"
Lloyd (1784-1859)

John Eager Howard (1752-1827)
governor of Maryland
m. Peggy Oswald Chew

Charles Howard (1802-1869)
——————————— m. Elizabeth Phoebe Key (1803-1897)

(1840-1909)   Charlton Hunt Morgan (1839-1912)
(1838-1916)   m. Ellen Key Howard (1840-1925) ——

John Pierpont
Morgan (1857-1913)

nt Morgan (1869-1935)   Ellen Key Howard Morgan (1873-1956)
ary Tinklepaugh

ghan Morgan (1910-   )   Isabel Merrick Morgan (1911-   )
William Scherp (1908-1974)   m. Joseph Dunlap Mountain (1900-1970)

Douglas Hunt   James Mountain
)   Whitaker (1934-   )   (son of Joseph Mountain
                                  and Audrey Daryll James)

1

LEXINGTON

One wouldn't be in such danger
From the wiles of a stranger
If one's own kin and kith
Were more fun to be with.
Ogden Nash

After [Morgan's] death . . . I took occasion to visit
his birthplace and ancestral home and found it most
interesting. It explained a great deal about his char-
acter.

Julian Huxley

IN 1933, at the climax of a life full of awards and honors, Thomas Hunt Morgan was given the Nobel Prize in "Medicine or Physiology" for his contributions to the chromosome theory of inheritance. He and his co-workers in the "Fly Room" at Columbia University had established the foundations of a new science called genetics, which was to revolutionize modern biology.

Morgan took this latest honor with the same noncha-lance as he took the honorary degrees stuffed in desk drawers and the elaborately inked awards fallen behind bureaus. His colleagues at the California Institute of Technology (where he had gone in 1928 to develop and head a new division of biology) heard about the Nobel Prize from the newspapermen who had come to Morgan's laboratory to interview him. His family read about the honor in the newspapers under the only picture he allowed the news photographers to take of him—one with whichever neighborhood children were standing about. Morgan did not even attend the official presentation in Stockholm, despite the case of prohibition whiskey given him by Caltech's board of trustees in order that he might practice for the festivities. He told the Nobel Committee he would come next year, combining the Scandinavian trip with visits to old friends and a little faculty recruitment for his biology department.

In April 1934, Morgan and his wife set out by train for the East Coast, from there to take passage to Europe on the *Majestic* with the youngest of their four grown children. In New York, Morgan stayed overnight with Dr. Warren Weaver. As Weaver remembered the evening, the father of modern genetics appeared at the door dressed typically in a rather disreputable old topcoat. In one pocket he had a comb, razor, and toothbrush wrapped in newspaper; in the other he carried a similarly wrapped pair of socks. "But what else do you need?" he asked the surprised Mrs. Weaver.

At sixty-seven Dr. Morgan still had dark hair, though it was turning gray. He was six feet tall and held himself straight. His eyes were a startling blue. He looked well—as he always did, since he was almost never sick (except for a peptic ulcer which would cause his death eleven years later). And he came prepared to enjoy himself—as he always did, once he had been torn away from his work.

The Weavers brought out a specially saved brandy for the occasion. Dr. Morgan appreciatively cradled it in his arms like a baby and remarked on the appropriateness of the date. The Weavers asked, "You don't mean you were born in 1865?"

His answer was No, that he was born in 1866—but 1865 was the year he was laid down.

It was an auspicious year for a geneticist to be laid down. It was the same year that Mendel laid down the fundamental laws of inheritance. The reports of Mendel's breeding experiments with garden peas were published the year of Morgan's birth but were promptly lost to view, not to be rediscovered by biologists until 1900, when Morgan was already a professor of biology at Bryn Mawr College.

Eighteen sixty-five was important in another way to Morgan's life. It was the final year of the American Civil War, in which so many of his family had been involved. Several of the men who knew Morgan personally or with whom he himself talked at length have referred to his family as being of English Cavalier stock, an aristocracy both impressive enough and distant enough that Morgan felt comfortable with it. Morgan once described himself as having enough Welsh blood in him "to leaven the whole damn Anglo-Saxon lump." But back home in Kentucky most references to his family went right to the heart of the matter: the name Morgan belonged first and foremost to his uncle.

For example, in 1936 when Thomas Hunt Morgan was seventy years old, the University of Kentucky decided to honor its famous alumnus, then and still in 1976 the only Kentuckian to have won the Nobel Prize. He was the author of twenty-two books and about 370 papers. Biologists from all over the world traveled to see his laboratory, and such scientists as Albert Einstein sat down at his dinner table. He was the true father of modern genetics. Yet the *Lexington Herald-Leader* headline on September 25, 1936, announced the celebration plans in the only way that would have seemed appropriate to most of its readers: "DR. MORGAN, NEPHEW OF 'THUNDERBOLT OF CONFEDERACY,' TO BE HONORED TODAY."

The Thunderbolt of the Confederacy—or the King of the Horse Thieves, depending on one's side in the Civil War—was Brigadier General John Hunt Morgan, a handsome, gallant, daring, occasionally reckless soldier whose raids into Kentucky were some of the most important Confederate efforts in the state, and certainly the most dramatic. Kentucky was a Union state officially, but many

Kentuckians, especially in the richer Bluegrass area, had a soft spot in their hearts for the Confederacy. After the war, the state grew decidedly southern, and one of its best legends was John Hunt Morgan. Morgan leading his loyal men in the face of overwhelming odds. Morgan escaping from the Ohio federal penitentiary and then through Union territory to rejoin his men. Morgan thundering on horseback through the front door of the family home to kiss his beloved mother and then thundering out the back door minutes ahead of the Federal troops.

General Morgan was killed during the war, two years before Thomas Hunt Morgan's birth, but his name remained very much alive. No one did more to keep it so than Tom's father, Charlton.

The family began in Kentucky in 1795, when Thomas Hunt Morgan's great-grandfather John Wesley Hunt migrated from Trenton, New Jersey, to Lexington. There he set up a small store and trading route and developed them into what is thought to be central Kentucky's first million-dollar fortune. In 1814 he built Hopemont, the gracious house where Tom would be born. The building still stands on the corner of Second and Mill streets in Lexington, maintained as a memorial shared, somewhat unequally, between Morgan the soldier and Morgan the scientist.

John Wesley Hunt's daughter Henrietta married Calvin C. Morgan, a businessman from Huntsville, Alabama, whom she—and her father's business concerns—eventually persuaded back to the Bluegrass. On a large farm on Tates Creek Road, at what was then the outskirts of the small town of Lexington, now opposite Mount Tabor Road, the Morgans raised a family of six sons and two daughters (a third died in infancy) in a uniquely southern form of aristocratic poverty. There was no ready cash, but life and education were conducted with expectations of the patriarch's will which, when read soon after his death in 1849, left Henrietta the family place Hopemont and established the first Morgan family of Lexington in its rightful place in society.

Tom's father, Charlton Hunt Morgan, was the fourth son, fifteen years younger than the eldest brother John Hunt. Charlton was good-looking, intelligent, and ambitious. At twenty he graduated from Transylvania University, a few minutes' walk from Hopemont. Instead of entering the thriving hemp and trading business run by his older brothers, he accepted a position as United States consul to Messina. Arriving in Sicily in 1859 at the start of the revolution, he unhesitatingly threw himself on the side of the nationalists. He became the first consul to recognize the Garibaldi government and—although he kept his title as American consul—fought as Garibaldi's aide-de-camp and was wounded.

As that war ended, the American Civil War began. Charlton returned home to serve as a captain in John Hunt Morgan's cavalry. He accompanied Brother Johnny on most of the 1862 and 1863 raids. He was wounded once and captured three times, to be held prisoner before being traded for a Union officer of equal rank.

All the Morgan boys rode with John Hunt, but only one, other than the general himself, was killed. That was nineteen-year-old Thomas Hunt Morgan. He had been the first to join the Confederate Army, and he had fought recklessly and joyously, seeming to delight in exposing himself to danger. He was captured, imprisoned, exchanged, and then hurried back to action. In July 1863, during a brief skirmish in Lebanon, Kentucky, John Hunt ordered him to leave the front because of his reckless behavior. But when the charge was sounded, Lieutenant Thomas Hunt Morgan pushed ahead and was shot through the heart. Charlton wrote home to his mother from the battlefield: "In the death of Tom I feel as my future happiness was forever blighted. I loved him more than any of my brothers."

Later that same month, John Hunt Morgan and several hundred of his men were captured in an overambitious raid that took them up through Ohio. Morgan and some of his brothers, including Charlton, had their beards and hair shorn and were confined in the Ohio State Penitentiary in Columbus. Morgan's escape, engineered by his brothers and the other officers, made up one of the chief tales of the future geneticist's childhood and was told to his own children, although by his mother and sister, not by Tom Morgan himself. It was a classic escape: a surreptitiously dug tunnel, a sprint past prison dogs and cruel guards, a dangerous railroad and horseback journey through federally held country to join his men again.

John Hunt resumed his raids, even if less successfully. Never the strongest of strategists, he now seemed to lose control of his men. There was a bank robbery, for example, in which the Raiders were said to be implicated. Meanwhile, Charlton and his brothers remained in prison, severely restricted as punishment for their part in the escape. They were miserable and none more so than Charlton. Every day their mother wrote highly censored letters which were delivered irregularly, and Charlton carried on an increasingly intense and affectionate correspondence with Miss Ellen Key Howard, a third cousin from Baltimore.

In March 1864, almost a year after his capture, Charlton was transferred to Fort Delaware, a regular military prison camp. He was still a prisoner when news came of John Hunt Morgan's death in Greeneville, Tennessee, on September 4, 1864. Only the following February—after nearly two years of imprisonment—was Charlton finally released. By the time he arrived in Virginia where his brother-in-law Basil Duke had assumed Morgan's command, Lee had surrendered at Appomattox and it was time for everyone to go home.

On December 7, 1865, Charlton and Ellen Key Howard, "Nellie," were married in front of several hundred of Baltimore's most distinguished citizens gathered in the Emmanuel Episcopal Church. Ellen Key Howard was the granddaughter, on her mother's side, of

Francis Scott Key of Star-Spangled Banner fame. Her paternal grandfather was John Eager Howard, Revolutionary War hero and from 1788 to 1791 governor of Maryland. She was often described as "lovely as a primrose" and, more important, as a true daughter of the South. Her romance with Charlton had been sweetened by his sacrifices for the Confederacy, and this shared devotion remained a major link throughout their marriage.

The new couple returned to Hopemont. In Lexington, the Morgan family fortunes had fallen with the loss of civil and property rights for those who had aided the Confederacy. The hemp industry in Kentucky was also flagging, and the Morgan brothers had not made the timely switch to tobacco. Crowded into the family home with the elder Mrs. Morgan were: her son Calvin, now, after John Hunt's death, the oldest son and chief businessman of what was left of the hemp and trading company; Calvin's wife and his mother-in-law, a refugee from war-torn Richmond, Virginia; Richard Curd, his mother's special protector, not to marry until long after her death; Francis Key, the somewhat spoiled baby of the family, a veteran of both battle and federal prison camp at twenty—and now Charlton and Nellie, who was pregnant by the time she arrived in Lexington.

The baby was born in Hopemont on September 25, 1866. Charlton only hoped the baby would be "as brave and noble a boy as was his namesake." He wrote John Hunt Morgan's widow that he had intended to name the child John but then "thought that the General's name and memory would be perpetuated by history, and that Tom's would not, as his humble position when he fell, would not entitle him to a page in history."

And so Thomas Hunt Morgan, future scientist and Nobel laureate, was born into a grand old house rich with family and legend. He spent his first years there while his grandmother prepared a small gift for Charlton, Nellie, and the baby: a house directly behind Hopemont on

Broadway, now 210 North Broadway. Charlton supervised the building, but his mother paid for it and retained title. When Tom was four years old, his brother, Charlton, was born, and when he was seven his sister, Ellen Key Howard Morgan, completed the family.

The Morgans, like the Howards whom they visited in Maryland each summer and often in between, depending on Charlton's increasingly desperate efforts to find appointments in Washington, were a proud family, clinging to the habits and attitudes of the southern aristocracy long after they had lost most of their money. The women were religious, or at the least devoted to their respective Episcopal churches—the Morgans to Christ Church in Lexington, the Howards to Emmanuel in Baltimore. (When Thomas Hunt Morgan, the too-daring uncle for whom young Tom was named, joined a Baptist church before his death, the church records showed that none of his family was Christian, but this must have been more a judgment on the Episcopal church than on the Morgans.)

But the cause that consumed most of Charlton and Nellie's energies was always the lost one. As Charlton had less and less luck in getting political appointments, despite hard work and efforts to use family influence, he spent more and more time writing to old comrades and helping organize reunions of Morgan's Raiders. Hundreds of old soldiers attended these events and continued to do so even after the turn of the century. The first such ceremony during young Tom Morgan's lifetime took place when he was a year and a half old and the bodies of General John Hunt Morgan and Lieutenant Thomas Hunt Morgan were brought from their first burial sites to be reburied in the Lexington family plot. An almost equally dramatic ceremony of which Charlton was a key organizer took place when Tom was a college sophomore and hundreds of former Confederate soldiers rode noisily into town to stage an encampment. The general's only living child, a posthumous daughter named Johnnie, then nineteen, reviewed their military drills and was honored

8

with gifts in her father's name. Tom did not come home for the third major event honoring his uncle, the last organized by his father. He was then in his forties and beginning the research that would lead to the Nobel Prize. While the nephew examined fruit flies under a small hand-held lens in a laboratory in New York, much of Lexington turned out for the unveiling of a large equestrian statue of the Thunderbolt of the Confederacy, still standing in front of the courthouse on Main Street.

Throughout all Tom's childhood and adolescence, the name John Hunt Morgan echoed about him. When he came home from school in the afternoons, there was often an old soldier waiting in the parlor to see Captain Charlton and Miss Nellie, both of whom were always ready to help a former Confederate and his family. And publicly, there were not only the celebrations and the books and songs about his uncle, but any anniversary or the death of any of the literally thousands of men who claimed to have ridden with Morgan called forth a reminiscence or some other published retelling of the exploits and legends of Morgan's Raiders. Later, as Morgan's men grew scarcer, the death of a man who claimed to have shod Morgan's horse would be sufficient. These accounts increasingly overshadowed the old debate about whether Morgan had been a true hero of the Confederacy—or a military washout and perhaps something of a scoundrel, both suggestions to which the Morgan family were naturally sensitive.

But as consuming an interest as the past was to his parents, the past took up no time whatsoever in Thomas Hunt Morgan's life once he had begun his work. Even when he was a boy it seems to have been of little day-to-day import. From his earliest years Tom pursued his own interests. The Morgans and the Howards had never seen anything like him in either family. He went about looking "torn-down," and was considered something of a bookworm. He had a butterfly net! He organized his Lexington playmates and his Baltimore cousins into collecting expeditions on the outskirts of Lexington or in the mountains near the Howard summer home in Oakland, Maryland. His scientific technique wanted refinement; once when he and his cousin John Hunt Morgan began dissecting a cat the animal awoke, angrily leaped off the table and fled. When Tom was about ten he was given two rooms in the attic of the house on Broadway. He painted and papered them and here put his collection of carefully labeled stuffed birds, birds' eggs, butterflies, fossils, stones, and whatever else in nature caught his eye. These rooms were his special province, where things were left alone by other members of the family to the extent that the rooms were said still to have held the collections when the last of the three Morgan children, Tom's sister Nellie, who had lived in the house all her life, died in 1956. The family lines of Hunt, Morgan, Key, and Howard boasted businessmen and diplomats, lawyers and soldiers, but in all its carefully kept pedigrees there were no scientists. In today's genetical terms, which Morgan helped establish, perhaps he was a new mutation.

In 1880, a week after his fourteenth birthday, Tom enrolled in the preparatory division of the newly established State College of Kentucky, located in Lexington. This was the latest in a long line of institutional reorganizations, absorptions, and divisions that had characterized all higher education in the Kentucky Bluegrass region. Although State College was to survive and eventually become the University of Kentucky, Tom came in during its most piecemeal, chaotic years.

When the school had divided from Kentucky University two years earlier, it lost all of its property and buildings. In 1880 most of the student body of 234 and the faculty of 17 were crowded into a rented building in what is now Woodland Park. Every room from cellar to attic was occupied. Three-quarters of a mile away toward town, an additional three rooms rented from the Masons housed the commercial department, the chemistry department, and the education school. Lexington had donated its old fairgrounds for a new campus, and the present administration building was being built. But even two years later in 1882, when Tom entered the freshman class of the college program, accommodations were painfully inadequate.

The all-male student body was a rough and often rowdy bunch, to the mixed delight and disapproval of the townspeople and the newspaper-reading public. They lived under a strict regimen; all students, including Morgan, were military cadets, required to wear uniforms that cost twenty dollars (tuition was only fifteen) and to stand military drill an hour a day, five days a week. Reveille began the day at 5:30 A.M.; taps and lights-out ended it at 10 P.M. In between, the bugle sent boys to class, to compulsory daily chapel, to study hall, and to table. Furthermore, 189 regulations were carefully spelled out, and the faculty given the right, indeed the duty, to think of more. All students were compelled to attend "at least one divine service" every Sunday. Students were not permitted to carry guns or bowie knives, although many did anyway. The college president had to give individual permission for any student to keep newspapers or books other than texts in his college room. Little wonder that Morgan received a few demerits, most for being late for chapel and for disorder in the hall and classes.

The academic program was limited. Regular college students (as opposed to those preparing for commerce or teaching) chose between the classics and the sciences, and Morgan took the latter. Courses offered science majors included mathematics, physics and astronomy, chemistry, agriculture and horticulture, veterinary science, civil history and political economy, mental and moral philosophy, Latin and either French or German, practical mechanics, English, engineering, and landscape gardening.

But for Morgan, the heart of the curriculum was the four-year stretch of natural history, taught by Professor A.

R. Crandall, a slender and sharply bearded man, who had formerly been with the federal geological survey and was completing his doctoral degree while teaching all natural history and many of the other science courses at Kentucky. He was an excellent naturalist. Tom liked him; he later said he had never met "a finer character or a better teacher." As was true in most American schools, natural history was largely systematic botany, especially comparison and classification. Zoology was covered in the same way, but more briefly. The natural history sequence also included a course in health and human physiology, and some work in geology and geography, especially as these pertained to coal. Kentucky, already planning for its agricultural experiment station, also gave what emphasis it could to agriculture in courses devoted to cultivation and propagation, the laws of growth, and the relation of forests to agriculture.

Through Crandall's influence, Tom spent the summers of his college years in Maryland and Kentucky, working for the federal geological survey. The hot, dusty field-work pursuing coal and other mineral deposits, along with the boring chemical analysis that followed, convinced him he was not meant to be a geologist. The experience did provide a good anecdote for later life, which he related in his best Kentucky style. As a sixteen-year-old college freshman Morgan found himself standing before a small potbellied stove in a country store in the distant mountains with a room full of suspicious mountaineers wondering aloud about the government badge one of them had seen. As far as they were concerned only revenue agents carried badges, and revenue agents were just the kind of men who would claim to have come so far back in the hills to look for something as unimportant as coal. Tom broke the tension and proved himself no revenue agent by asking a fiddler in the room to play the sailor's hornpipe and dancing through the complicated jig as long as it took to change the atmosphere from hostile to friendly.

What science Tom didn't study under Crandall, he took under Dr. Robert Peter, the aging former dean of the medical faculty at the old Transylvania University. Dr. Peter was a legend of a man—physician, historian, excellent botanist, organizer of the first geological survey in Kentucky, a pioneer of science in the Ohio Valley. But all that had been a long time before.

In later life, Morgan would graciously speak of "the really sound instruction that we received" at the State College of Kentucky, despite admittedly primitive conditions, and the "surprisingly excellent group of teachers" that gathered there. But while Morgan was in school in Kentucky the excellent faculty were saying something quite different. Dr. Peter had objected to another faculty member's appointment to expand the college's feeble program in chemistry, zoology, botany, and veterinary science. The head of the college, President James Patterson, felt and said that Dr. Peter had not kept abreast of new developments in chemistry and that his lectures were fifty years behind the times. Furthermore, he was half-deaf and half-blind and his classrooms were constantly in a ruckus because he couldn't control the students. The argument began during Morgan's stay at the college and ended in 1887, just before he graduated, by Peter's being kicked upstairs as "Emeritus Professor." He could keep his laboratory but he was to stay out of the classroom.

The year after Tom graduated, Professor Crandall unburdened himself of his own doubts about the success and management of the science program at the college. Unfortunately, the friend he chose to confide in was a newspaper reporter. The paper in the nearby small town of Winchester reported that the college was badly managed, especially the scientific department. The state legislature of 1889–1890 called for an investigation and heard an acrimonious exchange between Crandall, who stood by his charge that President Patterson completely neglected the sciences, and Patterson, who retorted that he didn't but that the accusation was typical of Crandall, "a half-educated Yankee." Patterson grudgingly agreed that Crandall had some good qualities as a teacher, but in general the president felt the science department would have been much better off without him.

Besides the limited curriculum and the frictions between the science faculty and the administration, Morgan experienced problems of his own in college. Being the nephew of the Thunderbolt of the Confederacy was fine if your teachers or classmates were sympathizers, but could have disadvantages if they weren't. Tom's French professor, for example, was a former Union soldier who had been forced by Morgan's Raiders to ride a mule backwards the ninety miles from Cincinnati to Lexington. He is said to have almost failed Tom, out of spite for his uncle.

Dr. Peter himself also had some unpleasant recollections of John Hunt Morgan. Morgan had attended Transylvania University when Peter was on the faculty there. John Hunt was too high-spirited and eager for adventure to last it out, but the college dropout always revered his former teacher. Unfortunately, when the Civil War began, Dr. Peter was a staunch Unionist and cut the Raider pointedly. And since Dr. Peter was the surgeon-in-charge of the Federal military hospital in Lexington, General Morgan felt it necessary to have him placed in custody whenever the Confederates temporarily gained control of the town. Nonetheless Dr. Peter liked Tom, who was a good friend of his son, and he once set the two boys to doing an experiment on selection among sugar beets—work then considered chemical or agricultural, today classified as genetic.

Thomas Hunt Morgan received the only bachelor of science degree given in 1886 by the State College of Kentucky. He was elected valedictorian by a five to four vote of the faculty. William Prewitt, who received the other four votes, was salutatorian. The rest of the 1886 graduating class consisted of Robert Prewitt.

What did a man with a bachelor of science degree do in Lexington? Morgan didn't know. He once said he went to graduate school because he didn't want to go into business and didn't know what else to do; he said elsewhere that he chose Johns Hopkins University because Joseph Kastle, a former science student at State College of Kentucky, had gone to Hopkins two years earlier. Furthermore, Baltimore was the ancestral home of the Howards, his mother's people, and the Howards (his mother included) doubtless felt a Baltimore school was appropriate. "But little did I know then," Morgan continued, "how little they appreciated that a great university had started in their midst, and I think this was typical of most of the old families in that delightful city." It hardly mattered. Whether Morgan realized it at the time or whether it actually was simple good luck, Johns Hopkins University was a most appropriate place for a biologist.

2

HOPKINS

*The future of the world lies in the hands of those who
are able to carry the interpretation of nature a step
further than their predecessors; . . . the highest
function of a university is to seek out those men,
cherish them, and give their ability to serve their kind
full play.*

Thomas H. Huxley
*Inaugural Address at
Johns Hopkins University*

WHILE WAITING FOR Johns Hopkins University's fall
semester to begin in 1886, nineteen-year-old Tom
Morgan began his study of marine biology at a small
summer school located on an inlet of Ipswich Bay, Massachusetts. Individual instruction, at a dollar a week, was
keyed to the teaching or research interests of the thirteen
men and thirteen women enrolled. Since the school at
Annisquam was to become the nucleus of the Marine
Biological Laboratory at Woods Hole the following year,
and since the MBL was a cooperative effort that involved
Hopkins biology faculty, Morgan may well have been
steered to Annisquam to learn the basic laboratory tech-
niques he needed for graduate work in biology. He found
the work rewarding and fun; he wrote back to Kentucky
that he continually congratulated himself on being there
rather than on the geological survey again.

Whatever Morgan was expecting of Hopkins itself, he
must have been equally delighted and self-congratula-

tory. In 1886 the university was ten years old, its academic programs well established, and its reputation secure in the educational circles of America and Europe. It was carefully designed to be different from State College of Kentucky and from most other colleges and universities in America.

Privately and generously endowed, Hopkins had none of the usual ties and obligations to religious denominations or state and regional politics or to the curriculum requirements spelled out for land-grant colleges by the federal Morrill Act. What it did have was an imaginative president, Daniel Coit Gilman, and a board of trustees who took seriously the Hopkins motto, The Truth Shall Make You Free. Consequently, on the evening it officially opened, Hopkins had dealt with and settled the conflict over Darwinian evolution in the schools, an issue many universities would have to face forty years later. Hopkins did so by accident. England's leading biologist, Thomas Huxley, was invited to give the opening address. Huxley was a well-known supporter of Darwin's theories of evolution, a man who dubbed himself "Darwin's bulldog," and to many people this mild-mannered (and not unreligious) scientist—who had come to talk about education—stood for all that was materialistic and irreligious and threatening about the science of biology in general. Furthermore, the university board of trustees, though a sincerely religious group, had decided a prayer would be inappropriate for a university lecture and had omitted to open with one. As one minister wrote, "It was bad enough to invite Huxley. It were better to have asked God to be present. It would have been absurd to ask them both." By the time Morgan arrived in Baltimore, the resulting hullabaloo had been forgotten. A strongly Darwinian faculty sometimes gave lectures on the subject to the Baltimore community, and students and faculty were sought after as adornments for the famed social life of the city. What remained from the occasion was a climate of free interchange of ideas.

Hopkins was one of the few institutions at this time to emphasize graduate study rather than undergraduate education. Hopkins fellowships, larger and more numerous than at any other institution, drew superior students from all over the country. Most of these were already college graduates like Morgan, who studied a limited number of subjects.

Of most importance to Morgan, Hopkins paid a great deal of attention to biology. Except for Harvard, American colleges barely touched on this subject. In fact much of science was frequently a stepchild in American curricula, where philosophy, history, and literature held priority. Or, as at the State College of Kentucky, science was tied to practical purposes such as agriculture or the geological surveys. And in fact, part of the reason for the original strength of biology at Hopkins, especially the inclusion of so much physiology, was the university's preparation for the 1893 establishment of the medical college and hospital specified in the original Hopkins bequest. But the two strong biology teachers recruited by Gilman set about building a department in which biological study and research had a value of their own, quite apart from potential use to medical students or immediate use to anyone. In the ten years between opening in 1876 and Morgan's arrival in 1886, a new biological laboratory had been build on campus and a marine biological laboratory set up on Chesapeake Bay with a station at Beaufort, North Carolina, and one in the Bahamas. The biology department had even established its own journal, *Studies from the Biological Laboratory at the Johns Hopkins University.*

For Morgan, coming to Hopkins was like coming into a new family, one whose existence he had scarcely been aware of. Studying under the men in the small biology department meant placing himself in the direct intellectual line of the greatest names in biology. Morgan divided most of his time between physiology and morphology, the two basic subjects offered. The chairman of the department, with whom he studied general biology as well as physiology, was Dr. Henry Newell Martin. Scottish-born and Cambridge-educated, Dr. Martin had studied physiology under Michael Foster and biology under Thomas Huxley, with whom he had coauthored a widely used textbook in elementary biology. Morgan's chief instructor in morphology was a stout, casually dressed, tobacco-chewing American named Dr. William Keith Brooks, former student of Louis and Alexander Agassiz, the naturalist and zoologist father and son at Harvard University. Brooks had participated in the Agassiz's efforts to establish marine laboratories for biological research and was also an alumnus of the famous marine biological station at Naples. During Morgan's stay at Hopkins yet other faculty came, bringing their own heritage and influence, and a series of distinguished lecturers visited the university. Many of Morgan's fellow students—notably E. G. Conklin and R. G. Harrison—were stimulating in school and remained lifelong friends and colleagues. Morgan acquired a special guardian angel in the Hopkins alumnus Edmund Beecher Wilson, ten years his senior and already a professor at Bryn Mawr College.

The circle to which Morgan had been admitted was not simply Hopkins, of course, but biology in general. There were not many trained scientists in America at the end of the nineteenth century, although the number was to increase rapidly. The early sons and daughters of the "family" knew and helped each other, sometimes divided on questions of theory and method but united in what they saw as a need to improve American biology's standing in the world and to change its direction.

The biology faculty during Hopkins's first twenty years are sometimes credited with training the generation of American zoologists who did just that. The traditional nineteenth-century approach to biology was descriptive: a scientist observed how an organism was put together, what its form and structure were, and—following Linnaeus in the century before—how it fitted into the great classification of species. The new biologists added a new level of inquiry. They wanted to know how the living organism worked. "Vital processes cannot be observed in dead bodies," the chairman of the biology department at Hopkins said firmly. Martin was not an exciting teacher, but his insistence on experimentation and his belief in the value of physiology in biological research were effective and stimulating.

Laboratory training had not gained universal respect in American higher education at the time Hopkins opened, and the faculty tried to strengthen it with standards and a clearly expressed philosophy. Naturally, the sciences were leaders in this new approach to education, but even history professors sometimes referred to their seminars as "laboratories." Graduate students in biology received almost all their training in the laboratory, where they were supervised daily. There were virtually no traditional recitations or lecture courses, although the faculty did suggest reading lists and the library bought the books individual students required for their work.

Beginning students concentrated on practical work to gain an acquaintance with the methods and instruments employed in biological research. Before beginning any original investigation, the student repeated some important research recently published and verified or criticized it. Thus Hopkins taught its young scientists that no man's work was sacrosanct, even that of the Hopkins faculty. Brooks, like Martin, although by nature an observer and philosopher rather than an experimenter, believed strongly that no theory should be accepted as dogma but only as the beginning for further investigation. He included Darwinism in this.

Hopkins also introduced Morgan to his lifelong distrust of fancy equipment. Martin wanted students to do hard and solid work, and he saw the practice of repeating experiments as a way the department had of protecting itself from "triflers" who had "a burning desire to undertake forthwith complicated research . . . believing . . . that laboratories are stocked with automatic apparatus—some sort of physiological sausage-machines, in which you put an animal at one end, turn the handle, and get a valuable discovery at the other."

Once the student passed the test and was elevated from trifler to serious scientist, the department tended to suggest a problem that needed experimental investigation, leaving the student to work out his own method of attack and apparatus, and then criticizing the results. Brooks once explained his neglect of Morgan's fellow student Conklin on the basis of his own strong belief in Darwin's law of natural selection. He felt it was a kindness to students to let them find out whether they had it in themselves to go ahead with their own research work without outside assistance. He had, in fact, given Conklin disastrous advice concerning a topic for his dissertation research, but in many, perhaps most instances— including Conklin's, who went on to become a fine biologist—the method was successful.

The experimental approach to science found at Hopkins was not a completely new idea in the late nineteenth century, as some of its proponents occasionally seemed to argue. Scientists in the sixteenth and seventeenth centuries had already begun to question the writings of the Greeks, accepted as the final word for centuries. Galen had written in the first century A.D. that there was a bone in the heart; Vesalius in the sixteenth century finally dissected a human heart and found none (and had the daring to say so). Moreover, in the seventeenth century, experiment proved that the heart pumps blood. The physician who proved it, William Harvey, also applied experimental methods to marine animals—and concluded that almost all animals are derived from eggs and that both parents contribute equally to the new organism—an idea that overturned Aristotle's notion that the female merely nourishes the male-produced embryo. The effect of examining for oneself the processes of life was liberating and productive. In 1839 the cell was proved to be the basic unit of living things. At about the time Morgan entered college, a series of important and related discoveries began in Europe, made mainly by cytologists (scientists who had become specialized—that in itself a nineteenth-century development—in the study of the cell). The testicular origin of sperm was established (many investigators were astonished to discover it did not arise in the blood, a theory that still lingers in phrases like "full-blooded Indian"). The fusion of one egg and one sperm was observed under the microscope, and this was recognized as fertilization. Chromosomes, dark-staining bodies, were discovered in the nucleus of cells in 1875, and in 1887 it was noted that the fertilized egg of the roundworm derived half its chromosomes from each parent. In short, during Morgan's college and early graduate education, scientists realized for the first time what was required for fertilization and were ready for experimentation to discover exactly what role the mysterious chromosomes played in the development of the resulting embryo.

Morgan was quickly convinced by the experimental approach at Hopkins. It was compatible with his own hardheadedness and common sense and with his desire to prove things for himself. He embraced the teachings of his professors in this area wholeheartedly and later even faulted Brooks for being too philosophical and not sufficiently experimental.

During his first years at Hopkins, Morgan began the first of his own experimental work, forming the principles he would express in *Experimental Zoology* (1907).

The essence of the experimental method consists in requiring that every suggestion (or hypothesis) be put to the test of experiment before it is admitted to a scientific status.

We demand in the case of a problem in experimental science that the conditions under which an event takes place be discovered, and that, if possible, we reproduce artificially the result by controlling the conditions. In fact, the control of natural phenomena is the goal of experimental work. . . . The investigator must . . . cultivate also a skeptical state of mind toward all hypotheses—especially his own—and be ready to abandon them the moment the evidence points the other way.

The ability to reject what was false, that is, what was shown to be false by his own experiments, even when he himself had once held the contrary, was to give Morgan a flexibility that few scientists possessed. It was to allow him to make dreadful errors, then correct them, and return to the forefront of thinking.

Morgan was moving toward the subject matter he would deal with all his life. In his early work in embryology and regeneration he was fingering tentatively a large question: How are cells regulated and controlled? In varying forms this question was to grip Morgan all his life:

The growth of animals and plants offers a wide field for experimental study. Under certain conditions we see a young animal continuing to grow larger until a certain size is reached, when growth slowly ceases. Although the animal may live for many years longer, it has ceased to grow. What makes it grow? Why does it stop growing? . . . We say it dies a natural death, and this seems inevitable, but only because we have found that death always takes place under ordinary conditions. Suppose, however, we change the conditions; might we not hope to prolong the duration of life?

Why, on average, in most animals, are there equal numbers of two forms—male and female? Is there an external mechanism? If so, what regulates it? Do external or internal conditions determine that one egg becomes male, another female? Even if an internal mechanism exists, it might be affected by external conditions and in any case the cause of the production of the two types must be determined.

Later in the same book, Morgan wrote, "The most distinctive problem of zoölogical work is the change in form that animals undergo, both in the course of their development from the egg (embryology) and in their development in time (evolution)." Certainly Morgan was frequently exposed to the theories of evolution at Hopkins, especially working so closely with Brooks. The Hopkins faculty *were* Darwinian—not only the biologists but most other professors too—and Brooks was Darwin's advocate in the university and in the community. In 1883 he had published his most famous and widely read book, *The Law of Heredity*, and dedicated it to Darwin, who had been buried in Westminster Abbey the year before. In his classes, in long and philosophic discussions with students, and in the Chesapeake Zoological Laboratory and its various stations, Brooks bent much of his energies toward generating interest in heredity. During the summers of 1883 and 1884 he worked with William Bateson, later to become the geneticist who introduced Mendel's laws both to English science and to English medicine and who did early and important studies on Mendelian genetics. The English geneticist acknowledged Brooks as the first man to intimate that "there was a special physiology of heredity capable of independent study." Morgan resisted, at least temporarily, this intimation, although his studies in regeneration led him to study adaptation and then natural selection, thus preparing him to enter the arena where Darwin's theories were still being debated. Morgan somewhat distrusted Brooks's philosophic stance, and this may have helped to influence him toward embryology and away from evolution, for which he always professed a distaste; yet Morgan did manage to write five books on the subject.

If Morgan had indeed come to graduate school because, as he had said, he didn't know what else to do, then at Hopkins he found out. The first year of study he led his class in biology and was an eager and excited student. After two years of immersion in biology, he was altogether the professional man. He had studied at the Chesapeake Zoological Laboratory and its branches and had participated in a scientific expedition to the Bahamas. His first article had appeared in the small department journal edited by Martin. Morgan's two-page contribution reported the conditions under which chitin solvents dissolved the horny material surrounding common cockroach eggs. Other articles were in preparation—some notes on the breeding habits and embryology of frogs; the growth and metamorphosis of tornaria, the larval form of the sea-acorn *Balanoglossus;* and a description of the dance of the lady crab. These and others would soon appear in *American Naturalist, Popular Scientific Monthly* (then more scientific than popular), and the *Journal of Morphology.* Much of this work was descriptive, but it reflected the solid training in morphology and physiology and the scientific methodology that would enable Morgan to conduct the experiments to concern him the rest of his life.

Having spent two years at Hopkins, Morgan was in 1888 eligible for a master of science degree at the State College of Kentucky, where the graduate program was simple enough: two years of study at another institution followed by examination by the Kentucky faculty to see if the student met Kentucky's standards. Morgan's old faculty went further. By unanimous vote they offered him a full professorship.

His State College classmate who had preceded him to Hopkins, Joseph Kastle, had returned to Lexington as professor of general, organic, and agricultural chemistry—and remained there to develop the agricultural experiment station. Perhaps his return was something like what Morgan originally had intended. Certainly the Kentucky school was so sure of Morgan's answer that they listed him in the 1888-1889 catalogue as Thomas Hunt Morgan, M.S., Professor of Natural History. If he had gone back, he would have been the replacement for Crandall, who had resigned in disgruntlement with the science program. But now Morgan had different plans. He would stay in school.

Money was fairly short at home. His father was again without a regular job although, since his efforts to find any kind of political appointment continued to fail, he had signed on as a life insurance agent. Tom's mother's health was delicate, partly from asthma and partly, perhaps, from fashion, since she took the cure at a nearby health resort and lived to a good old age. His sister Nellie was beginning the preparatory course at State College. His brother Charlton was uncertain what he should do. At twenty-two Morgan was the only member of the family with a reasonable skill to sell. Fortunately he didn't have to do it. He was awarded one of the highly competitive fellowships that did so much to assure the quality of the student body during Hopkins's early years. The standard award was $500 a year, almost as much as some of the younger faculty made, although financial setbacks in 1888 caused the trustees to ask fellows to pay their own tuition, thus effectively reducing the award to $400. Morgan said this meant he would live for yet another year on a student's frugal budget, but he leaped at the opportunity to continue his research projects and complete his doctorate rather than devote his strength to the large undergraduate teaching loads at a state college and to the building up of a fledgling department, in somewhat of a shambles from the Crandall and Peter affairs. He had gone to Hopkins a naturalist (and natural history remained an emphasis all his life, sometimes to the astonishment of more specialized scientists), but at Hopkins he had discovered experimental biology. When he referred to "his work" in a polite letter of refusal written to State College of Kentucky President Patterson, Morgan didn't even define the phrase. In his mind it loomed so large and pressing that there seemed no room for confusion. "My work is all in front of me and to stop now might mean never to go further which you as well as I would I think regret."

Shortly after penning this letter in Baltimore, Morgan was on a steamer to Boston, and from there he traveled some seventy miles southeast by train to Wood's Holl, later Woods Hole, Massachusetts. A small and remote seaside village that had once been a whaling center, Woods Hole was now the site of the Marine Biological Laboratory, established the year before to continue and extend the teaching and research program at Annisquam, where Morgan had gone in 1886. At Woods Hole, warm water from the Gulf Stream converged with cold water from the Gulf of Maine and the Labrador Current. In their meeting washed a remarkable abundance and diversity of marine organisms, and eventually a no less remarkable collection of biologists gathered there from colleges and universities all over America.

The MBL, like earlier marine stations founded by Agassiz and other American biologists, was the direct descendant of the great zoological station in Naples, founded by the German zoologist Anton Dohrn in 1872. A visit to the Naples station always inspired American biologists with energy to try for the same thing in America—a place near the ocean where biologists of all interests could come together to work, free of academic pressures and near a convenient source of varied marine organisms. It would be several long years before the little center at Woods Hole came anywhere near this. For the first four years it consisted only of a laboratory in a simple wooden building on a 78- by 120-foot lot. Fourteen years passed before the station got together enough money to buy its own ocean frontage. During these years it used the frontage—and some of the laboratories and equipment and personnel—assembled at Woods Hole in 1885 by the United States Bureau of Fisheries. The MBL also used some of the buildings left empty by the death of the local whaling industry. From the beginning, however, Brooks at Hopkins helped organize the MBL and the university itself had supported the laboratory and kept "tables," that is, subscriptions, to allow a certain number of its students or faculty to participate in the program.

It may well be that the establishment of marine biology stations was a vital factor in the development of biological research in America and in the changing direction of studies of biology. At a time when descriptive and structure-oriented biology still prevailed—as in the State College of Kentucky's natural history curriculum, and as symbolized to many scientists by the formal museums filled with collected, classified, and dead materials—waterside stations like Woods Hole came to represent a functional approach to the living organism. Under the direction of Professor Charles O. Whitman, a biologist at the University of Chicago in the winter, students at the Marine Biological Station became acquainted with the organism as it functioned in nature, and then brought it back to the laboratory, usually still living, to study it under controlled conditions—with a gradually more and more physiological and experimental approach.

In the first years, however, interest was still predominantly morphological. Even so, the interest was in the total creature, so that a new acquaintance at the station would not ask about one's special field of study but "What's your beast?" (And in the claims over beasts, the jealousy that did not extend to personalities or research problems existed in full measure.) Morgan's was the sea spider. Brooks, who had taken Morgan to Woods Hole and was directing his dissertation research, suggested that he attempt to define the phylogenetic relations of the sea spiders, *Pycnogonidia;* that is, to show exactly where they fitted into the Linnaean classification of species. Linnaeus himself had first raised the question, with the standard answer provided by Anton Dohrn who said that they were crustaceans like lobsters, not spiders. Morgan used the novel approach of studying their embryology, which proved that they were properly classified as true spiders. Morgan presented this work in 1890, in the first series of weekly lectures held at Woods Hole.[1] His dissertation on the subject met with Brooks's extreme approval; it was accepted for publication (by the journal *Studies from the Biological Laboratory of the Johns Hopkins University),* thus fulfilling a necessary condition

28

for completion of the doctorate. It was 76 pages long, with 8 illustrative plates, and was said to have nearly bankrupted the journal.

Social needs and pragmatism led to patronage of a common dining room, called the Mess, in a private home. The plain food was both good and cheap, board running five dollars a week until World War I prices forced it up to seven dollars. A local person did the actual cooking, but the waiters and waitresses were usually students, and young Morgan had his turn at helping with the organization of meals and housekeeping (probably for the only time in his life).

Woods Hole provided more traditional fun as well. Swimming was the principal sport. Since Tom enjoyed swimming and exercise, he doubtless joined in the game of leaping off the abandoned wharf of the defunct Pacific Guano Company (another failed industry at Woods Hole) into the eighteen-foot water. The isolation of the wharf meant that swimmers could do without suits.

The Fisheries Commission and the MBL organized baseball games, and tennis tournaments were held later. Any collection of young people invites horseplay—duels with water-filled paper bags and college-boy pranks with live lobsters. And as more and more students began coming to Woods Hole, it became known as a branch office of that heaven in which marriages are said to be made, and later as the Institute of Practical Eugenics. One of the Morgan children believes it served that function with their parents—that Edmund B. Wilson once dragged Morgan, who was reluctant to leave his work, out of the laboratory at Woods Hole, in order to introduce him to one of Wilson's brightest students at Bryn Mawr, a young woman named Lilian Vaughan Sampson, interested in embryology and, from that moment, in Thomas Hunt Morgan.

In the spring of 1890, Morgan received his doctorate from Hopkins and won the Bruce Fellowship in Research, a fairly new award for which regular Hopkins fellows competed. This enabled him to travel, conducting more research in the waters around Jamaica and the Bahamas and making a trip to Europe where he remained long enough to marvel at the Naples zoological station.

He spent the summer of 1891 at Woods Hole and by the end of August was on the train to Boston. From there he traveled to Johns Hopkins to pick up his mail for the last time, and to find a ride to Lexington where his grandmother Henrietta Hunt Morgan was dying. She had shown her pride in Tom's success by leaving him a share in the house on Broadway—and her only portrait of John Hunt Morgan. Her own house, Hopemont, where Tom had been born, would be sold shortly after her death on September 7, 1891.

Later that month Morgan turned twenty-five. He had grown the beard that made him look older and more dignified. In the fall, Morgan moved from Johns Hopkins in Baltimore to its sister school, Bryn Mawr, near Philadelphia. He was to replace Edmund Wilson as associate professor of biology, while Wilson went on to Columbia University after a visit to the Naples zoological station and two years in Europe in preparation for his new role.

BRYN MAWR

*If the origin of life on the earth, its evolution culmin-
ating in the races of mankind, seems to call for a
miracle, then the same or at least another miracle
seems called for each time an egg develops into an
adult organism.*

Thomas Hunt Morgan

BRYN MAWR COLLEGE had been founded by Quakers
in 1885 to offer advanced education for women comparable to that "so freely offered to young men." And though
Morgan was leaving the totally male Hopkins for the
totally female (except for the faculty) Bryn Mawr, he
should have felt at home. Not only did he succeed his
fellow Hopkins alumnus, Wilson, but so many Hopkins
men were among the school's first trustees and teachers
that it was aptly labeled "the Miss Johns Hopkins" or
"Janes Hopkins." Its academic program was closely mod-
eled after Hopkins's. Certainly Bryn Mawr was as close as
women seeking graduate education in biology could get
to Hopkins itself (undergraduate women could go to
Goucher in Baltimore). The one woman admitted to the
Hopkins biological laboratory by Chairman Martin had
been promptly ordered out by President Gilman because
"the Biological Laboratory where experiments in respect
to animal life are in progress is not well adapted . . . to
the co-education of young women and young men."[1]

For the first time Morgan was faced with a classroom

and the necessity of being in it every day. Biology instruction for undergraduate and graduate students was divided between four staff members. Morgan was an associate professor of biology; Jacques Loeb, whom Morgan knew from Woods Hole, and with whom he would continue close professional association all his life, was an associate in physiology and physiological psychology. The other two staff members were in physiology and botany.

Morgan lectured five days a week, at least twice a day, on either general biology or general zoology and on advanced biology. In addition, he gave a weekly series of lectures on embryology. He was also responsible for the laboratory work in all these courses and the research and dissertations of doctoral students, and he held a fortnightly evening journal club to discuss recent biological literature.

The easiest way to manage this teaching load would have been to teach the same thing over and over, semester after semester—which Morgan did not do. He chose instead to vary the specifics of courses to reflect some of his own research. He did this particularly in graduate courses and particularly after his return from a year at Naples in 1894-1895.

As a teacher, Morgan's style differed greatly from Wilson's, who for six years at Bryn Mawr had delivered smooth, thoughtful, and highly organized lectures. Morgan's dictum, which he later urged on his best graduate students, was Neglect Your Teaching.

When some ten years later Morgan noticed Fernandus Payne, one of his graduate students in the biology laboratory at Columbia, showing a strong interest in teaching, he teased "Watch out, or you'll become a dean." Real work, as Morgan saw it, was always original research, not the classroom. He taught his classes in much the same way as he wrote many of his books—taking them as an opportunity to muse aloud, formulating his current thinking on issues and presenting the latest research.

Some students were slightly confused by the wealth of information—fact and theory—that poured ramblingly and spontaneously out of their teacher, who often arrived late to class and occasionally did not arrive at all. But others were hooked, both on Morgan and on biology as he presented it. He had the enthusiasm, combined with the knowledge, to make an interested student aware of the wide world of biology and of the fascinating and varied methods of experimentation. Moreover, he was very generous with his time; he could always be interrupted for a discussion of work. Thus he easily acquired a faithful following who missed him when he went on leave: "The new man is nice, but he never serves us tea in the lab."

During Morgan's first three years at Bryn Mawr he had fun. He was a familiar sight on the college tennis courts, and he went to most of the college functions where, if he didn't often dance, he did at least noticeably do his duty by the refreshments.

During this same time the relationships in the Morgan family fell into a basic pattern that remained throughout Morgan's life. He became the special protector of his mother and sister, both of whom he loved dearly, but he did not go to them—they came to see him. He became a more detached son and brother to the two Charltons, with whom he had had little in common for some time.

Neither of his siblings had inherited their older brother's intellect, ambition, or perhaps his good luck. Nellie, after failing the entrance exam for Bryn Mawr in 1891, passed the following year and was placed under Morgan's wing. While determined to make her independent-minded, he himself treated her quite paternally. She must learn to think for herself, he lectured gently, but at the railway station she must wait hidden in the ladies room. Nellie's studies at Bryn Mawr were erratic, then postponed because of poor health, and never completed. Partly responsible for this change in plans may have been Nellie's lack of enthusiasm for the life of the mind and her attachment to her captivating mother, with whom she

33

would always remain, despite either Tom's plans for her or the hopes of a long list of suitors.

During Nellie's brief stay at Bryn Mawr, her mother often visited her and Tom—while her younger son, a dropout from the commercial department of the State College of Kentucky, wrote daily letters to his father ("and give my love to the old lady") from a dull office job at the Missouri Pacific Railway Company in Saint Louis. He begged his father to use his influence to get him a job at home *as he promised,* seemingly unaware that Charlton Sr.'s influence had seldom succeeded in getting himself a job.

This familial dichotomy, less apparent after Morgan married and his wife took care of all the correspondence, was at its strongest in 1893 when Morgan's father wrote a new will. He acknowledged that his own mother's rather chastising will had left him without any maternal inheritance or say-so over the family property, but nevertheless he wanted anything he had to go to his son Charlton, who "has always felt for me an affection and love, nearer and more tender, than that of my other children."

As a young teacher, Morgan first continued his research along the lines of his graduate school projects, largely descriptive studies on the *Balanoglossus* or sea acorn, Amphibia such as frogs, and the ascidian worms. Since he was concerned with embryological studies, marine organisms were an obvious choice: they were small, plentiful, and translucent. But the thrust of his research quickly changed. During this period Morgan became even more convinced of the power of the experimental method to which he would genuflect all his life. When he won the Nobel Prize forty years later, his first comment was that it was an honor for *experimental* biology—as opposed, he was implying, to the merely morphological and descriptive.

To the influence of the Hopkins faculty, of Jacques Loeb at Bryn Mawr, and of C. O. Whitman at Woods Hole was added that of the laboratories of the Stazione Zoologica, located under Mount Vesuvius by the Bay of Naples. Morgan had visited this mecca for biologists in 1890; in 1894-1895 he was given a year's leave of absence from the classrooms of Bryn Mawr to work there.

After the usual tourist and educational stops, Morgan arrived in Naples in October, his pockets filled with Italian money and a grammar book in hand. The Italians are said to have welcomed him with open arms because of his father's support of Garibaldi, but their gratitude was insufficient to influence the American government to give Charlton the consulship for which he had once again applied. Morgan in turn responded to the Italian people and declared he intended to marry an Italian princess, but he made no effort to locate one. What was happening in the laboratory at Naples was far more engaging. In an admiring article for *Science* (3[1896]:16–18) Morgan described the station as a kaleidoscope that changed from month to month, with students, professors, and researchers from all over the world, each with his or her own approach and interests. "No one can fail to be impressed and learn much in the clash of thought and criticism that must be present when such diverse elements come together."

As for Morgan, who was there to work with Hans Driesch and several other German biologists, he became interested in the experimental study of development. Development is a process of duplication and differentiation. Adults, whether adult frogs or adult humans, are formed from a single cell, the fertilized ovum, that duplicates itself some forty or fifty times. The basic question was (and largely still is) this: when a cell splits itself into identical halves, and each daughter splits itself into identical halves, and so on fifty times, all in an identical environment, how is the mysterious transformation made into separate and distinct sections of the body, some daughter cells becoming bone, some blood, and some brain?

Morgan set about investigating several internal and

external influences in the development of the egg, using simple Rube Goldberg apparatus. For example, he investigated the influence of gravity on the development of the egg with the following method: "A water-motor furnished the power. A bicycle was turned upside down; the rubber tire removed from the front wheel, and a string, running from the wheel of the motor around the rim of the bicycle-wheel, caused the latter slowly to rotate. The wheel made from twelve to sixteen revolutions per minute. The eggs were put into large test-tubes, closed at one end with a cork. These tubes were fastened to the spokes of the bicycle-wheel. . . . The tubes were nearly filled with water, but a large bubble of air was left at the top. As the tubes revolved the bubble of air would pass from one end of the tube to the other end, causing the water and the eggs to swirl twice during each revolution."

The most important contemporary debate was between the preformation and epigenetic theories. It was said by some that the cells were destined—or predetermined—or preformed—to turn into a complete creature. According to this theory, called the mosaic or preformation doctrine, within every ovum there was something like a tiny manikin that would automatically unfold and grow into an adult, rather like the clever Japanese pellets that, in water, grow into strange flowers. In the alternative theory of "epigenesis," development was thought to depend on forces that interacted with the material stuff of the egg, and complex interactions between protoplasm and environment and nucleus were assumed to be possible throughout development.

If everything was preformed, development would proceed on rigid lines and there would be nothing for the zoologist to do except to see how the history of the race was repeated in the development of the individual. The epigenetic theory, on the other hand, was dynamic, and despite the disadvantage to Morgan of having to admit the possibility of vital forces, it more readily allowed the development process to be studied experimentally.

In 1883 the embryologist Roux proposed that the nucleus divides qualitatively to produce a mosaiclike distribution of potentials in the daughter cells. Thus the first division of the zygote separates the cells that will become the right-hand side of the body from those on the left; and the future development of all parts of the body is determined in the first few cell divisions of the egg. Roux supported his proposal five years later with the discovery that when one of the two daughter cells of a frog's egg was killed, the single remaining cell developed into an incomplete embryo, with one-half missing. The issue seemed settled; but then in 1891 the respected embryologist Hans Driesch performed a similar experiment and discovered that a single daughter cell of a sea urchin egg did not form a half embryo, but developed normally. Clearly if one-half or even one-eighth of the bundle of cells has the same potential as the whole original ovum— and even has the ability to recognize that circumstances have altered—then there must exist a more complex interaction between the protoplasm and total environment than the preformation theory allowed. For some years there was confusion because of the contradictory findings of different workers. For example, Chabry and Conklin found mosaic development for ascidian eggs, while in 1893 Wilson confirmed Driesch's remarkable discovery with *Amphioxus* (a very advanced worm or a very primitive vertebrate). Morgan, working with Driesch at Naples, and afterwards on his own at Bryn Mawr, sought to disprove the mosaic theory by stimulating the ovum and embryo in a variety of ways. In 1895 with the white fish *Fundulus* he succeeded in demonstrating that a single daughter cell would develop normally. This line of work was extended in some delicate and beautiful work by Hans Spemann who produced Siamese twins of varying degrees by tying a hair around the egg in early development.

After Morgan confirmed Driesch's discovery that a half-

or a quarter-embryo could grow into a complete adult, he went on to investigate the forces that made this happen. Taking the two-celled embryo of *Fundulus heteroclitus*, he found, as Chabry had, that destroying one cell might result in an incomplete adult, but when the remaining cell was turned upside down, or shaken, or centrifuged, normal development was restored and a complete adult was produced.

During this period of his life, Morgan was also setting the pattern of later work in which he was to experiment, using eclectic methodology, with at least fifty different creatures. And he was easily and agreeably diverted into byways whenever the experiment seemed interesting. For example, he made a series of experiments attempting to produce embryos showing spina bifida, a defect of the vertebrae caused by a failure of the neural ridges to unite. He found that the condition could be produced by two entirely different methods, either by the addition of a weak salt solution to the embryo or by direct and specific injury. Spina bifida occurs in man, but Morgan would not have seen his own work as having relevant application, for it was not fashionable then to regard man as an animal.

In studies on sea urchin eggs, both Morgan and Hertwig independently discovered that the ovum could be artificially provoked to begin dividing by the addition of either magnesium chloride or hypertonic sea water. Ordinarily the process of dividing would require a sperm. The work was completed by Morgan's old Bryn Mawr and Woods Hole colleague Jacques Loeb, who in 1899 succeeded in inducing the ovum to develop artificially by using a minor modification of the salt solutions that Morgan had been using. This work was to occupy a large part of Loeb's life and led to the fatherless production of normal larvae and later of tadpoles and yet later of both male and female frogs. The newspapers found this exciting and the publicity was quite helpful to the Marine Biology Laboratory where most of the work had been done. But there was also some feeling among the scientists at the laboratory that Loeb had received all of the credit. Morgan himself, however, bore little resentment and remained on good terms with Loeb.

Another experiment should be mentioned because it triggered Morgan's distrust of one of the great biologists of the day. Morgan repeated Boveri's experiments on fertilization of eggs from which the nucleus had been removed. Boveri had found the resulting sea urchin resembled the father, but when Morgan repeated the experiments the results were too variable to warrant this conclusion. After this he found Boveri's work less acceptable.

When Morgan returned to Bryn Mawr in 1895, he was promoted from associate to full professor of biology. In 1897 his first book was published—*The Development of the Frog's Egg: An Introduction to Experimental Embryology*. In this book he reviewed the results of experiments on the early embryogenesis of the frog, considering the determinants of the plane of cleavage, the production of abnormal embryos with spina bifida, the effects of injuring isolated daughter cells, and the development of the three basic tissue layers. Meiosis is roughly described, but no mention is made of heredity or of sex determination. Moreover Morgan expressed some distaste for Weismann's theoretical interpretation of meiosis: "Weismann has utilized the discovery of the reduction of the number of chromosomes to build up an elaborate and highly speculative theory of heredity. . . . The reduction of the chromosomes to half the number present in the other cells of the body, seems, according to Weismann and others, to be a preparation for fertilization. . . . the number of chromosomes will thus remain constant for the species from generation to generation."

Morgan came back from Naples convinced that Americans were isolated from much of the newer, more current work in biology and lacked a marine biology center of European quality. During the next few years Morgan visited several fledgling American marine biology stations, mostly on Cape Cod, but he kept coming back to the Marine Biological Laboratory at Woods Hole. The young station underwent an important change in 1897.

C. O. Whitman, the progressive-minded director of the laboratory, ran head on against the board of trustees, largely made up of the original Boston founders. Whitman argued for expansion; the more financially cautious board argued for not opening the laboratory at all that summer since there was no money to do so. There were, however, more than 300 scientist-members, their dollar-a-year dues fully paid. At the threat of having their marine organisms breed and reproduce without them, the members elected a new board of trustees—a bigger one, more nationally representative, almost all scientifically involved, and much more confident and daring about the Marine Biological Laboratory's future. The twenty-four trustees elected in August 1897 included Morgan. He remained on this board until 1937 and was an emeritus member until his death. Until World War II gas rationing prevented it, he spent at least a part of every summer afterward at Woods Hole and continued to participate in its affairs, which grew less rocky as the laboratory became better established.

Both at Woods Hole and at Bryn Mawr, Morgan began to emphasize regeneration in his teaching and research; this followed naturally from the studies of injury to a daughter cell, since the ability of half the embryo to grow into a whole one was a sort of regeneration or replacement. Some of Morgan's early graduate work had involved regenerative studies of the earthworm, and now he turned his attention to several other humble creatures who let him cut bits off and then obligingly grew them back on again.

Regeneration was a fact, but the purpose of much of Morgan's work during this period was to find out under what conditions it occurred and what factors contributed to its occurrence.

"When an animal reaches a size that is characteristic for the species," he wrote, "it ceases to grow, and it may appear that this happens because the cells of the body have lost the power of further growth. That the cessation of growth is not due to such a loss of power is shown by the ability of many animals to regenerate a lost part."

All animals with blood regenerate their red blood cells continuously throughout life. All animals replace gaps in the surface layer of skin throughout life; but if a limb or a head is cut off, some animals, such as worms or snails, can grow a new one. And some, such as mice and men, cannot. In an experiment to search for the forces controlling regeneration, the future Mrs. Morgan grafted pieces from the flatworm *Planaria* together and found that grafts could be made between pieces taken from all different parts of the body. If the match was perfect, no regeneration at the cut surface occurred. If the match was imperfect—that is, if there was raw tissue left exposed—then a new head or tail regenerated from that spot. If one of the pieces was turned upside down and then grafted, two heads would regenerate at the line of the graft, one for each worm.

Morgan himself found that when the jellyfish *Gonionemus vertens* was cut into pieces, the cut edges came together, fused, and assumed a bell shape. He also found that when the tip of the tail of a fish or the base of an earthworm was cut off, it regrew slowly. If more was cut off, it regrew more rapidly, and the rate of growth was not influenced by the amount of food the animal ate.

But how does the cut end "know" whether to grow a new head or a new tail? Morgan offered a tentative answer to this question when he explained his work and views on regeneration in a series of lectures at Columbia, published in 1901 as part of the Columbia University Biology Series. Morgan's *Regeneration* still makes excellent reading, according to Dr. Richard Goss. He remarks "how few of the questions Morgan wondered about have yet been answered."

From studies of grafting and regeneration in the tadpole, fish, and earthworm, Morgan postulated a graded distribution of tensions or materials from hydranth to stolen—or from head to tail—highest at the head end, lowest at the tail end. These tensions or materials are disturbed by the removal of a part. Perfect grafts restore the tension, alleviating the need, otherwise present, for regeneration to take place. Without a graft, the cells where the body was severed know to regenerate a head if the pressure behind them is decreasing, a tail if it is increasing.

This theory was clearly spelled out in papers in 1904 and 1905 and is acknowledged by Child as the origin of his own more biochemical theories of gradients. Morgan saw all of his work as an effort to answer the basic question of how an ovum turns into an adult. Though the gradient theory seemed to answer part of this question, it left much unanswered. For this reason, Morgan returned to regeneration studies whenever he had the opportunity. In fact he was working on regeneration of the brittle stars when he died in 1945.

Had Morgan remained at Bryn Mawr, he doubtless would have continued solid and respectable embryological and regeneration work. But in 1903 his friend Wilson asked him to come to New York and occupy the country's first professorship of experimental zoology. Along with the chair, which was Columbia's—and Wilson's—commitment to experimental biology, went the clear understanding that Morgan would teach fewer classes (almost no undergraduate ones) and do more research. Wilson once said modestly that his only contribution to the genetics work at Columbia was his discovery of Thomas Hunt Morgan. But actually there might have been no genetics work at Columbia had Thomas Hunt Morgan not discovered Wilson and the cytology that he was doing. The decision to go to the larger school was a turning point in Morgan's career.

In 1903, almost as part of his preparation for leaving Bryn Mawr, Morgan became engaged, following Wilson's lead. Morgan was thirty-six, to be thirty-seven in September, and this seems to have been his first romance. He had always maintained that a man should be established and on firm financial footing before marriage, and he must have felt this meant engagement as well, since his new fiancée and he had known each other for several years.

Lilian Vaughan Sampson had entered Bryn Mawr College in 1887 and was one of Edmund Wilson's first and best biology students. After she graduated in 1891, she studied biology and the violin in Zurich. In 1894 she received a master's degree from Bryn Mawr. At some point she spent almost a year in Arizona, staying with friends. Ordinarily, however, she lived in Germantown, Pennsylvania, with the grandparents who had raised her and spent whatever time she could at Bryn Mawr, where she continued to attend lectures and work in the laboratory on her own embryology experiments and as a demonstrator.

When she and Tom became engaged, Lilian was thirty-four. Orphaned at three, she had become extremely close to her sister Edith and was virtually orphaned again when Edith finished Bryn Mawr and married. Lilian's aunts and grandparents were delighted to see her blossom under Tom's affection, and Lilian quickly became a devoted member of the Morgan family. As she wrote Mrs. Morgan, "When I think of the great love and affection that there is between you and Tom it overwhelms me and makes me feel very humble, because I realize what a big place his wife must fill; the wife of a man who has such a love for his mother. I know that I do not begin to fully appreciate him yet, although I love and honor him more every day. I look to you to help me know how to make him more happy."

During that whole cold and drizzly spring, Morgan left his work long enough once a week to make the trip between Bryn Mawr and nearby Germantown where Lilian lived with her aunts, at first by train and later, when the weather warmed, on horseback. In April he bought Lilian a white diamond. "I have always been happy," she wrote her future mother-in-law, "but now, life seems so brimming over with joy and happiness that there is not time or room for anything else."

Since the wedding was not to be until June, Morgan "very wisely" told his fiancée she need not think about details until the middle of May. As for himself, Lilian wrote, he "like all men simply submits to the inevitable."

The wedding was small, mostly family, and the new couple were soon aboard the train on Morgan's first trip west, to the Marine Biological Station at Pacific Grove, California, and a biologists' honeymoon. In the fall they moved to New York, where students began enrolling for Morgan's classes ahead of time and Tom and Lilian spent their spare time together, searching for live things in the ponds and sorting out furniture in their rented house. Every morning after Lilian saw to the cook and attended to some housekeeping chores, she walked the few blocks to Schermerhorn Hall on the Columbia campus and spent an hour or two working with her husband in his laboratory.

Although the patterns of Morgan's life had not greatly changed—academics in the winter and Woods Hole in the summer—the substance of it had. He had found the perfect wife who would protect him from the more bothersome details of the world. He would never have to hammer a nail, or learn to drive a car, or pack a suitcase. Lilian would read his manuscripts, understanding and closely following the most intricate details of his research. A scientist herself, her priorities after marriage were always clearly defined: (1) Thomas Hunt Morgan; (2) the children, soon to follow; and then and only then (3) her own work in the laboratory. Because of his new wife, Morgan was able to become even more involved with his work. At the same time he had found the perfect department within which to work. And shortly before, there had appeared in Europe the two ideas upon which the yet embryonic field of genetics would rest. The first was the rediscovery of Mendel's work, made independently and almost simultaneously about 1900 by three scientists, one of whom was Hugo DeVries. The second was DeVries's theory that mutation was the way in which new species originated. In 1900 Morgan had visited DeVries's garden and laboratory at Hilversum, Holland, and had gone on to Naples where the talk at the marine laboratory must have been largely of Mendel. After a long period of settling in at Columbia, Morgan was drawn back to this most exciting of developments in biology and would try his own hand at inducing mutation.

4

THEORIES,
FACTS — AND FACTORS

Evolution is something that happens to populations,
and without a mathematical theory, connecting the
phenomena in populations with those in individuals,
there could be no clear thinking on the subject.

Sewall Wright

THE TWENTIETH CENTURY began explosively. For almost 2,000 years Christianity had been the main source of creativity; in the eighteenth century it was supplemented by the observation of nature. And then along came chemistry and physics and genetics. Radium, a new element that emitted unknown rays, was discovered; atoms were found to be made up of smaller bits; even energy was found to be made up of bits; and light became pliable. In 1839 it had been discovered that all living things were made up of bits, called cells, all of which had been begotten by other cells and so back for 5,000 million years. The way that cells made daughter cells in their own likeness was clarified in 1900 when three botanists discovered two papers written by Gregor Mendel in 1866.

The commonsense notion of inheritance suggests that when parents have opposing characteristics, if one is black and one white, for example, or one tall and one short, that the offspring will end up somewhere between, by a so-called blending inheritance. Gregor Mendel, a Moravian monk of the Augustinian order of Saint Thomas, showed the commonsense notion to be wrong, at least in the case of garden peas *(Pisum sativum)*.

Mendel was born Johann Mendel in 1822 (the year after Napoleon died), in Brünn (now Brno, Czechoslovakia) which was within earshot of Napoleon's guns at the Battle of Austerlitz and close to Prague, where Landsteiner was to discover the ABO blood groups. Mendel received a good grounding in mathematics and physics at Vienna University but he was, like Darwin, an academic failure and could not even obtain a teaching certificate. Mendel was a skillful and patient husbandman, however, eliciting from his peas—which he called his children— marvelous results that he carefully categorized and counted. He also practiced that disappearing art of keeping up with the literature, and at some risk bought Darwin's books, which were all on the Index Librorum Prohibitorum.

Mendel chose to cross pea plants that differed from each other in a clear-cut way; for example, he crossed tall with dwarf plants, plants bearing smooth seeds with plants bearing wrinkled seeds, white-seeded plants with those bearing grey-brown seeds, or plants that flowered along the stem with those that flowered only at the end of the stem. The experiments were remarkable—first because Mendel had the wit to think of them, second because he had the skill to manipulate the plants without accidental pollination by intruders, and third because he realized that the simple ratios of characteristics among the offspring must reflect some fundamental structural characteristic. He concluded that the ovule or the pollen must contain something (later to be called a gene) that remained separate and distinct in the hybrid. Thus, either the paternal or maternal characteristic would become apparent—*not* a blend of both. And in the next generation, both the maternal and paternal characteristics would

Parents	Tall-Tall x Dwarf-Dwarf
	(all genes are paired in the adult)

Offspring in	Tall-Dwarf x Tall-Dwarf
1st Generation	(all hybrids, all look tall)

Offspring in	Tall-Tall Tall-Dwarf Tall-Dwarf Dwarf-Dwarf
2d Generation	(3 classes appear in 1:2:1 ratio. Members
Produced by	of the first 2 classes look tall—though
Breeding	the 2d class is hybrid. Plants of the
1st Generation	class that inherited no tall gene are short.)
"Incestuously"	

Figure 1. Mendel's principle of segregation

appear, and would appear in fixed ratios. In Figure 1 the
gene that manifests itself as the physical characteristic is
shown in boldface. Mendel noted that "transitional forms
were not observed in any experiment," and that dominant
(in this example, tall) and recessive (here, dwarf) forms
appear in the ratio of 3 to 1. Moreover, "the offspring of
the hybrids separated in each generation in the ratio of
2:1:1 into hybrid and constant forms. . . . If A be taken
as denoting one of the two constant characters, for instance the dominant, a the recessive, and Aa, the hybrid
form, in which both are conjoined, the expression $A +
2Aa + a$ shows the terms in the series for the progeny of
the hybrids of the two differentiating characters."

In a second series of experiments Mendel crossed two
characters in each parent to see if tallness goes with
smooth seeds or if the two characteristics would separate
and segregate independently, which in fact they did.
Here he had a remarkable stroke of luck. Peas have
exactly seven pairs of chromosomes and each of the seven
characters that Mendel chose to cross was located on its
own chromosome. He never knew this, of course, for
chromosomes had not yet been discovered.

Later he studied the results of crossing peas with three
pairs of alternate characters. In each case he found every
possible combination of characteristics in the offspring.

Any two pairs of alternate characteristics would appear in the ratio of 9:3:3:1. He inferred from this that all types of seeds were produced with equal frequency and, because of the dominance of some characteristics, produced offspring in this curious but predictable ratio. It is perhaps easier to understand the possible combinations and the ratio by using a square designed by Punnett (Figure 2).

Pollen types		TR	TW	DR	DW
	TR	TR TR	TW TR	DR TR	DW TR
	TW	TR TW	TW TW	DR TW	DW TW
Ovule types	DR	TR DR	TW DR	DR DR	DW DR
	DW	TR DW	TW DW	DR DW	DW DW

(The 16 possible types of offspring would sort themselves into the ratio of 9 tall round, 3 tall wrinkled, 3 dwarf round, and 1 dwarf wrinkled.)

Figure 2. Gamete (seed) types produced by
Tall-Dwarf, Round-Wrinkled first generation hybrids

Mendel's findings are often summarized as (1) the law of segregation, which states that genes do not blend but remain separate to segregate in the ratio of 3 dominant to 1 recessive when both parents are hybrids, and (2) the law of independent assortment, which states that one pair of genes assorts, or segregates, independent of all others.

Unfortunately Mendel did not have the pleasure of seeing the enormous influence of his hobby. The Brünn Society for the Study of Natural Science to which the original papers were presented in 1865 listened politely but in uncomprehending silence. The published lectures, modestly entitled "Experiments in Plant Hybridization," went out to about 120 universities and scientific organizations, but none of the botanists at any of these centers ever wrote to Mendel or gave any indication of having read his work. He must have felt very alone.

Mendel did correspond with Professor Carl Nägeli, a respected botanist in Munich. But Nägeli gave him no encouragement and in one patronizing reply advised Mendel to switch from experiments with garden peas, a perfect subject for breeding experiments, to Nägeli's own favored hawkweed (*Hieracium*), such an unsuitable subject that the new results must have made Mendel doubt his earlier work. (Hawkweed was unsuitable because the seeds are usually purely maternal in origin and arise without either meiosis or fertilization.) The entire town of Brünn mourned Mendel's death in 1884 and scarcely anyone outside the town noticed it. Leoš Janáček played the organ at the funeral, and soon afterward the new abbot burnt all Mendel's unpublished papers. Because Mendel published his work only once, in an out-of-the-way journal, because it was avant-garde, and partly, too, because biologists did not understand mathematics (and because as a monk Mendel was not considered a scientist), his work was to lie unknown until 1900.

Upon the discovery of Mendel's papers in 1900, scientists immediately asked whether Mendel's rules were true. And if they were true, how general were they?[1] Morgan initially believed the rules to be true since they were solidly based on experiment. In 1903 he wrote: "The importance of Mendel's results and their wide applicability is apparent from the results in recent years. . . . the theoretical interpretation that Mendel put on his results is so simple that there can be little doubt that he has hit on the real explanation" (*Evolution and Adaptation*, pp. 284–85).

But various considerations led Morgan increasingly to doubt that Mendel's theories were true after all—to doubt that the genes remained separate in the hybrid and that they sorted out independently. One reason for his growing dissatisfaction with Mendelism came from his own experimental attempts to confirm Mendel's laws.

For example, when Morgan crossed white-bellied yellow-flanked house mice with the wild type, his results were erratic and indicated that the germ cells carried other colors.

Because he could not confirm Mendel's findings himself, Morgan had come by 1909 to feel quite strongly that Mendelian theory was being given more credence than it merited. In that year, at a famous meeting of the American Breeders Association (the first group that clearly recognized the truth of Mendel's laws) he caused some surprise by an almost bitter attack, if not on Mendel, at least on those who had wholeheartedly adopted his views.

In the modern interpretation of Mendelism, facts are being transformed into factors at a rapid rate. If one factor [the word then current for gene] will not explain the facts, then two are invoked; if two prove insufficient, three will sometimes work out. The superior jugglery sometimes necessary to account for the results may blind us, if taken too naively, to the commonplace that the results are so excellently "explained" because the explanation was invented to explain them. We work backwards from the facts to the factors, and then, presto! explain the facts by the very factors that we invented to account for them. . . . In the first place the assumption of separation of the factors in the gametes is a purely preformation idea . . . while the epigenetic conception, although laborious, and uncertain, has, I believe, one great advantage, it keeps open the door for further examination and re-examination. Scientific advance has most often taken place in this way. . . . There is a consideration here of capital importance. The egg need not contain the *characters* of the adult, nor need the sperm. Each contains a particular material which *in the course of the development produces* in some unknown way the character of the adult.

During this period between 1904 and 1910 Morgan was developing his own theory of alternating dominance to explain hybrid behavior.

I think that the condition of two alternative characters may equally well be imagined as the outcome of alternative states of

51

stability. . . . In one and the same individual both the dominant and the recessive characters may appear. For example, chocolate and black mice give Mendelian results; but I have individuals that were black in front and chocolate behind. Again black eye and pink eye are Mendelian alternatives, yet I have three mice with one pink eye and one black one. Local conditions, I infer, determine in these heterozygotes that at one time the dominant and at another the recessive character come to the front, and I could bring forward evidence to show that the results are not due to segregation of unit characters [genes].

Although Mendel's theory was losing its appeal to Morgan, over the same period he became increasingly satisfied with the mutation theory. In 1900 Morgan visited Europe and in particular Hilversum, Holland, where the botanist Hugo DeVries had come across Mendel's papers and in 1886 discovered some evening primroses, *Oenothera lamarckiana*, growing in a waste field near his home. They had the remarkable property of producing new types that bred true—DeVries called them mutations. Out of a total of 50,000 individuals he later obtained about 800 mutants that fell into seven types. If the primrose can suddenly produce new types, even new species, DeVries reasoned, this might be a visible example of the way that new species in general were produced.

DeVries explained mutation as an alteration of a distinct material unit that by 1915 was considered an organic chemical molecule. At least it was so considered by Morgan: "It is difficult to resist the fascinating assumption that the gene is constant because it represents an organic chemical entity."

Clearly if new races or species can arise spontaneously from large mutations that occur frequently enough, then no other theory is needed to explain the diversity of living things. There were, however, at least three such theories then widely promulgated. The oldest—that God created each species intact in a single and spectacular six-day

52

week in 4004 B.C.—was already under heavy fire from supporters of Charles Darwin.

Darwin's theory of evolution appeared in *The Origin of Species* (1859), which sold out its first edition of 1,250 copies on the day of publication. "I view all beings not as special creations," Darwin wrote, "but as the lineal descendants of some few beings which lived long before the first bed of the Cambrian system was deposited. . . ."

It is interesting to contemplate a tangled bank, clothed with many plants of many kinds, with birds singing on the bushes, with various insects flitting about, and with worms crawling through the damp earth, and to reflect that these elaborately constructed forms, so different from each other, and dependent upon each other in so complex a manner, have all been produced by laws acting around us. These laws, taken in the largest sense, being Growth with Reproduction; Inheritance which is almost implied by reproduction; Variability from the indirect and direct action of the conditions of life, and from use and disuse: a Ratio of Increase so high as to lead to a Struggle for Life, and as a consequence to Natural Selection, entailing Divergence of Character and the Extinction of less-improved forms. Thus, from the war of nature, from famine and death, the most exalted object which we are capable of conceiving, namely, the production of the higher animals, directly follows. There is grandeur in this view of life, with its several powers, having been originally breathed by the Creator into a few forms or into one; and that, whilst this planet has gone cycling on according to the fixed law of gravity, from so simple a beginning endless forms most beautiful and most wonderful have been, and are being evolved.

Darwin's discovery of evolution occurred during his voyage around the world on the *Beagle*, 1831–1836, on which he assembled a huge collection of plants and animals and fossils. However, it was one thing to realize that evolution had occurred, quite another to explain how it occurred. It was on reading the Reverend Thomas Malthus's *Essay on Population* that he discovered part of the explanation: "It is the doctrine of Malthus applied with manifold force to the whole animal and vegetable kingdoms: for in this case there can be no artificial increase in food, and no prudential restraint from marriage." As long as an excess of offspring is produced, only the fittest or best adapted individuals will survive in the inevitable struggle for food.

But Darwin ran into another difficulty: he knew that a giraffe with a longer neck would be better able to survive since he could reach more food, but it was not clear how the giraffe had come to be born with a longer neck and how the longer neck was inherited and preserved in the race. Darwin believed in blending inheritance, but this would tend to produce a balance between the parental extremes. How then could the race progress?

To answer this question Darwin reluctantly adopted parts of yet a third theory—Lamarck's inheritance of acquired characteristics. In 1809 Jean Baptiste Lamarck had proposed that the bodily skills, habits, or physical structures that parents acquire during their lifetime may be handed down to their children. Though Darwin wrote, "Heaven forfend me from Lamarck nonsense of a 'tendency to progression,' 'adaptations from the slow willing of animals' etc.!" he had already accepted it, according to his diaries, when Fleeming Jenkin, an engineer, pointed out that blending inheritance was mathematically incompatible with the gradual operation of natural selection because of the loss of half the genetic variability in each generation. Darwin could see no other way for the giraffe's neck to become long except that the stretching of one generation is inherited by the next.

These were the issues that were debated throughout the world. Morgan argued evolution at Hopkins, Bryn Mawr, Columbia, and Woods Hole. After dinner at one of the summer laboratories when Brooks, Morgan's old teacher, had been holding forth on heredity, Wilson once spoke up and said, "Brooks, I cannot see the logic of anything you say."

Brooks arose, let fly a stream of tobacco juice over the railing of the porch and replied, "I don't expect you to, Wilson. You have to think long and hard on these matters before they have any meaning to you."

But to Morgan and Wilson, Brooks's thinking seemed romantic and not the way to resolve the issue of evolution. It must be done through experiment, and that was the method Morgan decided to use, following the mechanistic philosophy of Huxley, Loeb, and the European empiricists, which was in 1903 forcibly expressed by Jules Poincaré: "Experiment is the sole source of truth. It alone can teach us something new; it alone can give us certainty."

Morgan decided that the evolution of an ovum into an adult could be studied experimentally and *should* be in order to disprove both Lamarck's theory of the inheritance of acquired characteristics and Darwin's theory of natural selection. (Before long Morgan would consider disproving Mendelism as well.) His own philosophical attitude at this time was elaborately and untidily set out in the first of his five books on the subject of evolution, *Evolution and Adaptation*. It was published in 1903, with a dedication "to Professor William Keith Brooks as a token of sincere admiration and respect." This was one of only three books that Morgan would dedicate (one went to his mother, the other to Wilson). Brooks was mentioned, somewhat negatively, once on page 463, but most of the 470 pages expressed Morgan's distaste for the theories that Brooks loved best—those of Darwinian evolution.

Evolution and Adaptation was chiefly an evaluation of the evidence for Darwinian evolution, although it also considered the case for the theories of Lamarck, DeVries, and Mendel. It concluded that Darwin and Lamarck were wrong, whereas DeVries and Mendel were correct. For example, Morgan did not accept the evidence for Darwin's struggle for existence. "Millions of individuals [bacteria] are present at the time when food supply becomes exhausted and they all pass into a protected resting stage." Fresh from his visit with DeVries, Morgan declared that "the time has come, I think, when we are beginning to see the process of evolution in a new light. Nature makes new species outright. Among these new species there will be some that manage to find a place where they may continue to exist. . . . Some of the new forms may be well-adapted to certain localities and will flourish there, others may eke out a precarious existence, because they do not find a place in which they are well-suited, and they cannot better adapt themselves to the conditions under which they live and there will be others that can find no place at all in which they can develop, and will not even be able to make a start. From this point of view the process of evolution appears in a more kindly light than when we imagine that success is only attained through the destruction of all rivals. The process appears not so much the destruction of a vast number of individuals, for the poorly adapted will not be able to make even a beginning. Evolution is not a war of all against all, but is largely the creation of new types for the unoccupied, or poorly occupied places in nature."

Morgan slowly came to accept evolution and natural selection over the next thirty years or so, but his discomfort with the mathematics remained a difficulty. While by the end of his career he did acknowledge the mathematical contributions of Haldane, Fisher, and Wright, Morgan still stuck by his original insistence that evolution be treated experimentally.

Another difficulty was that he (like most people) found it hard to admit that great developments can be due to tiny accidental events, particularly when he was not able to see the stages in the development. He changed his mind partly through the influence of his students at Columbia (soon colleagues) who continually argued, in H. J. Muller's words, that "Darwin's theory of natural selection was undoubtedly the most revolutionary theory of all times" and that "Darwin's masterly marshalling of the evidence for this . . . remains to this day an intellectual monument that is unsurpassed in the history of human thought." Morgan was always more impressed by concrete evidence than by intellectual monuments, and thus the issue may well have been resolved for him during his visit to Oxford on June 12, 1922. Julian Huxley arranged for the Hope Department of Zoology to get out a representative sample of examples of adaptive coloration in a number of groups of insects, including Poulton's wonderful series of butterflies illustrating mimicry—which hardly admits any explanation other than natural selection. Huxley described Morgan's reaction: "When I went back to fetch him for a luncheon, I could hardly prevail on him to move, 'This is extraordinary! I just didn't know things like this existed!' " Huxley described the occasion years later to the American Philosophical Society and concluded, "It was, I am proud to believe, the occasion of his conversion to a belief in adaptation and the efficacy of natural selection in producing it."

Morgan's various books on evolution showed an increased if still grudging acceptance of evolution in Darwinian terms. But even in his last book on the topic, although Morgan accepted the major tenets of Darwinian evolution (but not the "reckless" use to which some naturalists put it), he had not yet laid aside all reservations:

The implication in the theory of natural selection, that by selecting the more extreme individuals of the population, the next generation will be moved further in the same direction, is now known to be wrong. Neither the genetic factors responsible for a part of the initial variability, nor the environmental factors, can bring about such an advance. Without this postulate, natural selection is impotent to bring about evolution. On the other hand, if variations arise, owing to genetic factors (mutants) that transcend the original limits, they will supply natural selection with materials for actual progressive changes. There is here no implication that natural selection itself is responsible for the appearance of new types, some of which may have a survival value, except that owing to the destruction of the less well-adapted types room is left for the better adapted. If all the new mutant types that have ever appeared had survived and left offspring like themselves, we should find living today all the kinds of animals and plants now present, and countless others. This consideration shows that even without natural selection evolution might have taken place (*The Scientific Basis of Evolution*, pp. 130–31).

At least Huxley felt satisfied that Morgan believed and so happily he dedicated his 1942 book *Evolution: The Modern Synthesis* to "T. H. Morgan: Many-sided leader in biology's advance."

The side of himself that Morgan saw most clearly, even in the early years of the twentieth century was the embryologist. He later reiterated in *The Scientific Basis of Evolution* that the proof of evolution would be through embryology, not paleontology. It was as an embryologist that Morgan started and finished, at least in his own view. And leaving Bryn Mawr for Columbia, he continued his experimental work on embryological problems, many of which he summarized at this stage in his career in *Experimental Zoology* (1907), as follows:

He reported the experimental stimulation of the sperm of starfish by adding ammonia and prostatic fluid and salts; he reported the results of experimental crosses between different species of sea urchins and showed that survival was influenced by the season of the year and the temperature of the water. He was interested as always in proving epigenetic development and thus sought to diminish the supposed dominance of the nucleus and its chromosomes. "Most embryologists seem inclined to ascribe the effects entirely to the nucleus, which they believe dominates all the changes in the protoplasm. On the contrary, I am inclined to think that it has not been shown conclusively that this influence is nuclear in origin but may be due to the protoplasm introduced with the sperm." He quoted Driesch's findings that the method of cleavage of the embryo, its tempo, and all development in early stages exhibited the characteristics of the egg regardless of the kind of sperm used, facts which led Morgan to ask, "If chromatin in the nucleus is the all-controlling influence, why is there delay in the appearance of paternal elements?" In his struggle to escape the influence of the chromosome he even stated that "the sperm seems to be only chromatin, but there may be more cytoplasm that cannot be seen."

In a letter written to Professor Carl Nägeli on September 27, 1870, Mendel had suggested that sex determination might prove to be a phenomenon of heredity and segregation. Like Mendel's other ideas this was either not understood or simply dismissed. Even when Mendel's work was rediscovered in 1900, the connection seemed unclear. As Morgan himself asked during this early period, "How can sex be caused by genes? Which would be dominant, male or female?"

Laying aside the complications of that question, however, the cytologists in particular began asking whether the chromosomes somehow determined sex, since most chromosomes come in identical pairs except for one pair in one sex. These odd chromosomes were called X and Y. To many—Wilson for example—it seemed likely that they were sex determining.

Nevertheless, some facts seemed to argue otherwise. The oyster has the ability to change its sex with the weather, and creatures like the silkworm can change the ratio between male and female progeny according to environmental factors. Both earthworms, which are always hermaphrodites, and gynandromorphs of any species, are male and female at the same time. Work done in England showed female moths and birds to be the heterogametic sex, that is, to have an X and a Y chromosome instead of two X chromosomes, while in the United States research, mostly on insects, showed the male to be the XY or heterogametic sex. To compound the diversity and thus the confusion, in a certain homopterous insect that has both an English and an American form, in the English form unfertilized eggs produce males; in the American form they produce females.

The simple notion that the sperm carries either an X or a Y chromosome and thus determines the sex was made more difficult to accept by the proven existence of natural or artificial parthenogenesis. How could the sperm be sex determining when Loeb's experiments were producing male and female frogs without any sperm at all?

Morgan became progressively more interested in the mechanism of sex determination. In 1903 he reviewed the current theories and in 1906 he began a seven-year study of the way *Phylloxera*, a small fly with a partiality for grapes, is able to produce both males and females parthenogenetically. Before 1910, the results seemed inconsistent with chromosomal determination of sex.

In his 1907 discussion in *Experimental Zoology* of internal factors of sex determination, Morgan gave much space to nonchromosomal determinants. He noted that Nettie Stevens and Wilson had cleared up some of the most important questions in Stevens's work on the beetle *Tenebrio* and Wilson's on the bug *Lygaeus furcicus,* both studies showing male sex to have an X and a Y chromosome. "It seems to me, however, that a simpler hypothesis may be formulated. The sex of the embryo is not laid down as such in the egg or sperm but may be determined later by the quantitative reaction resulting from the activity of the chromatin in the cells of the embryo." He seemed a little critical of Weismann's suggestion that the purpose of sexual reproduction is to induce variability. Morgan always so hated teleological explanations that he commented, "Variability may be as marked in non-sexual forms produced parthenogenetically as in sexually produced forms."

At this period Morgan began studying the mechanisms of self-sterility, following Castle's discovery at Harvard that the hermaphroditic ascidian *Ciona intestinalis* fails to fertilize itself. "I have carried out a series of experiments on this and other species in the hope of discovering to what this action is due," Morgan wrote. "It is not possible to make the egg receive its own sperm by immersing it in the blood or extracts of the ovary of another individual. Conversely, it is not possible to make the sperm enter its 'own' eggs by soaking it first in the blood or testis-extract of another individual." He found the results complicated and difficult to interpret, but kept trying throughout his life—and was still trying when he died.

Morgan always had dozens of experiments going, many of which came, as he expected, to a dead end. He often jokingly said that he did three kinds of experiments— those that were foolish, those that were damn foolish, and those that were worse than that. But one experiment during this period, a Lamarckian one, produced an unexpected dividend. In 1908 Morgan set a graduate student to work rearing *Drosophila* in the dark, hoping to produce flies whose eyes would atrophy from disuse and disappear in further generations. Fernandus Payne had worked before on a blind lizard from Cuba and the eyeless cave fish of Indiana, and so Morgan suggested that he try a related experiment for which he could collect some *Drosophila* that would be attracted to bananas placed on the laboratory window ledge. *Drosophila* were small flies called variously fruit flies, vinegar flies, pomace flies, and even banana flies, after the fruit which had carried the first *Drosophila* into North America. This idea of using *Drosophila* as a convenient experimental animal came from Castle at Harvard, whose student Woodward had bred these flies since 1900 to study the effect of inbreeding. The Payne study, despite sixty-nine generations of *Drosophila* which never saw the light of day, came to nothing. The sixty-ninth generation emerged momentarily dazed (at which point Payne called Morgan to come quickly, teasing him with what looked like successful results)—but the flies soon recovered and flew to the window as if nothing had ever happened.

But what *had* happened was that an almost ideal choice of experimental animal had been introduced to Morgan's laboratory at Columbia University. It had no modesty about breeding, which it did publicly, rapidly, and prolifically. It ate cheaply—a little mashed up and fermented banana. And since it was tiny, thousands and thousands could be housed in small milk jars along the walls of the tiny 23- by 27-foot Columbia laboratory soon to be called simply the Fly Room.

The second experiment that Morgan conducted on *Drosophila,* again with Payne, may have been stimulated by DeVries's suggestion at Cold Spring Harbor, New York, in 1904 that mutations might be artificially induced. "The rays of Roentgen and Curie which are able to penetrate into the interior of living cells," he had said, should "be used in an attempt to alter the hereditary particles in the germ cells." And for two years Morgan and Payne subjected flies to "the emanations of an X-ray machine, . . . radium, . . . wide ranges of temperature, salt, sugars, acids, alkalis, without any resulting mutation."

In 1910, when he received a visit from his old Bryn Mawr colleague Ross Harrison, Morgan swept his arm around the laboratory, indicating the rows and rows of flies in their bottles. "There's two years work wasted. I've been breeding these flies for all that time and have got nothing out of it."

5

COLUMBIA

The investigator must . . . cultivate also a skeptical state of mind toward all hypotheses—especially his own—and be ready to abandon them the moment the evidence points the other way.

Thomas Hunt Morgan
Experimental Embryology

Everything of importance has been said before by somebody who did not discover it.

Alfred North Whitehead

BATESON'S OFT-QUOTED dictum Treasure Your Exceptions is difficult advice. Exceptions are easy to treasure once recognized; the trick is to recognize them. Without knowing what to expect, perhaps expecting large mutations of the type DeVries found in Lamarck's primroses, *Drosophila* workers found it hard to recognize tiny changes in wing shape or eye color as mutations. To get an idea of how hard it is to spot mutations in a fly a quarter of an inch long, try to see how many you can spot on the greatly magnified *Drosophila* gynandromorph in Figure 3.[1] Certainly Woodward, who bred *Drosophila* for two years at Harvard, noticed no mutations; he recommended the flies to Castle, who bred them for five years and found none; Castle recommended them to Lutz, who found at least one; and Lutz recommended them to Morgan, who was throwing up his hands in despair two years later.

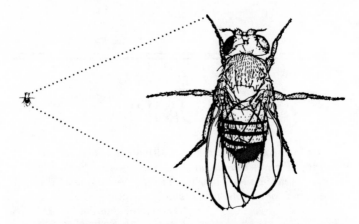

Figure 3. This *Drosophila* shows a number of mutations that are quite easy to spot with the help of magnification but not so easy when the fly is life-size. The left side is bigger than the right, causing the fly to bend to the right and indicating that, as females are bigger than males, the left side is female and the right side male. Such gynandromorphs occur roughly once in 2,200 *Drosophila* (Morgan's figure). The left side shows the dominant character notch-wing (from the mother), and the right side shows the sex-linked recessive characters scute, echinus eye, ruby eye, broad wing, and forked bristles, all of which entered from the father. Reprinted, by permission, from Morgan, Bridges, and Sturtevant, *The Genetics of Drosophila*, figure 49

(Payne, Morgan's student who first used *Drosophila* in the Morgan laboratory, could hardly have been expected to spot mutants in the dark.)

But in Morgan's laboratory, probably in May 1910, a male *Drosophila* was born with white eyes, instead of the red eyes of his brothers and sisters. This was clearly a mutant, and destined to become the most famous insect in scientific history.

Where did it come from? It is possible that Morgan actually had succeeded in inducing mutation among his fly stock. He reported in *Science* (1911) that he had subjected a number of flies, pupae, larvae, and eggs of *Drosophila* to radium rays the same month the fly was born. And in March 1911 he wrote his friend Loeb, "As I told you last summer, all my wing mutants go back to my flies treated with radium, as do at least two of the eye mutants."

It is also possible that Morgan inherited the mutant. Frank E. Lutz, who was at the Carnegie Laboratory at Cold Spring Harbor from 1904 to 1909 and later at the American Museum of Natural History, claims responsibility for the white-eyed fly in his fascinating book *A Lot of Insects:*

Professor T. H. Morgan visited the Station and I told him that a white-eyed Drosophila had appeared in one of the pedigreed strains but that I was too busy with abnormal veins to attend to it. He took live descendants of this white-eyed "sport" and bred from them. Eventually he got the white eye back. This little story is no particular credit to me. If I had realized how valuable that white-eyed mutant was destined to be, I would not have been happy to give it away. However, it fell into good hands and, really, Drosophila melanogaster should be called Morgan's Heredity Fly.

Morgan was reluctant to accept this version. Lutz's book was reviewed in *The Journal of Heredity* in 1942 by a reader who took Lutz at his word, provoking an immediate if somewhat vague reply from Morgan. Morgan wrote that he had indeed asked Lutz for a culture of *Drosophila,* but (1) it did not include Lutz's white-eyed fly, which was already dead when found; (2) it did not include descendants of the white-eyed fly, because pair matings would have produced some white eyes in the next generation, which did not appear; and anyway (3) finding the white eye, later to prove one of the commonest mutant types, "was not so important as the use to which it was put."

Morgan had at least two strains of *Drosophila* in the Columbia laboratory—those from Lutz and those collected by Payne. He obviously would have preferred the idea that the stock producing the white-eyed fly had ancestry somewhere outside the open window of the lab. And in the family story, which Morgan never corrected (though he listed the first white-eyed fly as born in May), the first white-eyed male was born out of the blue shortly before Morgan's third child, who arrived on January 5, 1910. According to this anecdote, Morgan rushed to the hospital where his wife's first words were, "How is the white-eyed fly?"

The baby was fine, but the fly was feeble. Morgan is said to have carried it home at night to sleep in a jar by his bed, returning it to the laboratory during the day. There it mustered enough strength to mate with a normal red-eyed female before dying, leaving the mutated gene in what was to become a prodigious family line.

Whenever the fly appeared—probably in May 1910, when Morgan wrote to Bateson cautiously that his "pomace flies looked interesting"—it was soon mated. Ten days later 1,240 offspring appeared. Virtually all were red-eyed. In Mendelian terms red eye is dominant over white eye, and thus would be expected to appear in all members of the first generation (since each fly would receive a gene for red eye from the mother, who would have only genes for red eye). But strangely there were also three white-eyed males in Morgan's first generation.

Genetically this was almost impossible. It could more easily be explained by nondisjunction or by imperfect control in the laboratory, carelessness either in breeding or counting—or by Lutz's story.[2] If Lutz's white-eyed male had lived long enough to breed and one of its female offspring was in the culture given to Morgan, then this fly would have inherited one gene for white eye as well as one gene for red eye (the first from her odd father, the second from her normal red-eyed mother). This female would have red eyes, since the red gene would be dominant, but any of her sons that inherited the gene for white eye would have white eyes. An understanding of why that should be so was at least another generation of *Drosophila* away.

Morgan dismissed the three white-eyed males easily,

as "due evidently to further sporting," i.e. mutation, and announced that they would "in the present communication, be ignored." He also ignored them thirty years later when responding to the Lutz account, and in fact said all offspring were red-eyed. All later accounts have perpetuated this error.

The next generation of *Drosophila* was only another ten days in coming. When the white-eyed fly's offspring were mated to each other, the results were in accordance with Mendel: 3,470 red-eyed offspring, and 782 white-eyed. Roughly one out of four had inherited and manifested the recessive characteristic. In neither generation had there been blending—the differential parental characteristics had segregated nicely.

Morgan wrote up the results of the breeding experiment hastily, and on July 7, 1910, submitted it to *Science*, mistakes and all. (The manuscript copy of this famous paper is in the possession of C. D. Darlington.) This first fly paper heralded the beginning of what Sir Gavin de Beer has called "the magnificent ocean of experiments on *Drosophila* by T. H. Morgan and colleagues which established beyond any reasonable possibility of doubt that Mendel's principles of heredity were correct."

There was an anomaly in the results of this first experiment, however. According to Mendelian expectations, in the second generation one-quarter of the males should have shown the recessive characteristic and one-quarter of the females. Instead, Morgan found that half the males had red eyes and half had white, but that *none* of the females had inherited white eyes. All females had normal red eyes.

The reason white-eyed flies were mostly male was soon shown by further matings. When a white-eyed male was bred to a normal female, all the offspring were red-eyed; when a white-eyed female was mated to a normal male, half the offspring were white-eyed and these were all male. Clearly the factor—or gene as it was soon to be called—for white eye was unlike other Mendelian reces-

67

sives in that the outcome was influenced by the sex of the parents. Morgan gave a rather complicated explanation that was wrong in detail but correct in its conclusion: that the eye color gene (R) and the sex-determining factor (X) were combined or, as we would now say, linked. "The fact is that this R and X are combined, and have never existed apart." (See Figure 4.) By late 1911 both Morgan and Wilson realized that the same mechanism explained the mode of inheritance of hemophilia and color-blindness in man.

Morgan hesitated to conclude that the X factor was the X chromosome, even though he knew through Nettie Stevens's cytological work that the female *Drosophila* has two X chromosomes and the male but one. There were several reasons for his hesitation. To begin with, he did not like hypotheses, particularly those whose survival depended on subsidiary ones as he felt the chromosome theory did: Second, his results conflicted with the studies on moths and birds in England in which certain characteristics appeared most often in the female, implying that the female has only one X factor and the male two. Third, the postulate that development was determined by an undemocratic group of dark bodies or chromosomes seemed to him merely another name for the preformation theory—which left no room for environmental influences and the influence of the cytoplasm. But last of all, chromosomes simply seemed unlikely candidates to be controlling bodies. Sex determination was only one of their inconsistencies; equally confusing was the diversity of the number of chromosomes between species: Drosophila has 8 chromosomes (4 pairs), a goldfish 104, and a little Spanish butterfly 380. Dogs and chickens have 78, horses 64, and humans had 48 until 1956 when they lost 2 in a recount. Were this not confusing enough, the chromosomes are very small and disappear between cell divisions as rapidly and completely as Congress in recess.

Morgan expressed yet another difficulty: "Since the number of chromosomes is relatively small and the char-

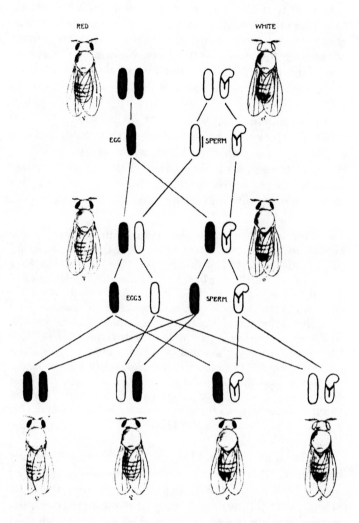

Figure 4. Cross between white-eyed male *Drosophila* and red-eyed female. The sex chromosomes are indicated by the rods. A black rod indicates that the chromosome carries the factor for red eye; the open rod indicates the factor for white eye. Reprinted from Morgan, *A Critique of the Theory of Evolution*, figure 58

acters of the individual are very numerous, it follows on the theory that many characters must be contained in the same chromosome. Consequently many characters must Mendelize together. Do the facts conform to this requisite of the hypothesis? It seems to me that they do not" (*American Naturalist*, 44:449–96; submitted May 8, 1910).[3]

And of course they did not. Not yet anyway.

Within a few months four additional eye-color mutants appeared; for example, pink eye, which segregated independent of sex and independent of the white eye, and vermilion eye which was "sex-limited" showing the same sex distribution as white eye.[4] He then conceded:

Moreover, it is clear why in one case (white and pink) there should be sex-limited inheritance and in the other (red and pink) a different kind of inheritance, provided, as the facts strongly indicate, that the factor for pink is contained in another part of the hereditary mechanism than the factor for white. In other words, the factor for white (absence of red) is connected with the factor that determines sex, while that for pink is contained in a different part of the cell. It is this evidence that has seemed to me to show that the phenomenon of sex-limited inheritance is due to an intimate physical relation between the sex factors and the other factors in question; and the most obvious connection is that the relation is to be found in the chromosomes that carry both the sex factor and those factors that are sex-limited (*Science* 33 [1911]:536).

In the same article Morgan showed himself utterly convinced that mutations in *Drosophila* followed Mendel: "One fact will especially impress itself on anyone who follows the history of these new types, the 'segregation' of the character and in most cases the absence of intergrades" (p. 496).

It soon was apparent that the genes were inherited together in groups corresponding with the number of chromosomes and therefore that the genes were most probably part of the chromosomes.

For all his time of not finding any mutations, Morgan now spotted one or two new ones each month. Actually, in the early months of 1910 before the white-eyed fly appeared, a fly was born with a dark trident-shaped pattern on its thorax. Another had an olive-colored body. A third had a beaded wing margin. Yet another showed unusual coloration at the base of the wing. But Morgan probably viewed these as merely insignificant birthmarks—not exceptions that he should treasure. After May, however, there were so many mutations that it is reasonable for us to ask whether Morgan did not really increase the mutation rate with the radium treatment. The question cannot be answered because the prevailing emphasis was on the size of particular mutations rather than on rigorous measurement of their frequency. Morgan himself was swept away in the more absorbing task of classifying and mating all the new types, and certainly his laboratory records were not designed to answer the mutation question. This can be seen from a letter he once wrote a fellow worker and former student: "My dear Mohr, I am sending off this brief statement concerning the stock that I sent you. It was a new truncate that came up in my cultures and I have even forgotten now in what culture, although I probably have a record of it somewhere."

By the end of 1912 there were forty types of flies that could be identified by visible abnormalities. Once a mutant was recognized, it was mated and its progeny mated, brother to sister, and then mated again with each parent, and with other mutants, in this way "constructing" flies with the desired genes for study. Morgan might have ordered up a culture of females carrying white eye as a marker gene on the first or X chromosome, with speck marking the second chromosome, olive body color marking the third chromosome, and bent wing marking the fourth chromosome. These females would be mated with a newly discovered mutant male and then Morgan could see with which of these various genes the newly visible gene of the male tended to associate. If it appeared with bent wing, for example, then the new mutant gene was clearly on the fourth chromosome.

The process involved the storage of thousands of flies, many of them kept in small half-pint glass milk bottles that Morgan "borrowed" from the Columbia University cafeteria. New generations were anesthetized with ether. The drugged flies were then spread out and counted under hand-held lenses or simple microscopes and either killed or, if further breeding experiments were planned, returned to their jars to recover and have another meal of fermented banana. Any experiment required the counting of thousands and thousands of flies, and during the peak of the *Drosophila* work a regular sight at the subway stop near Columbia was a group of students carrying home milk jars of flies to spread out on their own kitchen tables. When one of Morgan's student's children was asked what his father did for a living, he proudly said "My daddy counts flies for Columbia University!"

But however hard it was (1) to spot an abnormality in the first place, and (2) to correctly count the thousands of flies which either did or did not have it, the breeding experiments also demanded unusual brilliance to infer the invisible mechanisms that were producing the observed results. It was as though Morgan and his students were struggling to figure out the rules of bridge and the face value of cards in each concealed hand, merely by watching the game being played and arranging the hands deliberately, although without seeing the face value of the cards. In the same way they speeded up the process of observation and deduction of the rules of genetics by shifting genes from fly to fly.

Morgan's confidence that there was some relation between the "sex-limited" genes and the chromosomes that carried the sex factor was supported by additional data that he did not wish to divulge at the time of the first article—namely the discovery of miniature wing. This mutation also showed the same mode of inheritance as white eye, or was sex-linked. So now Morgan had three factors (white eye, vermilion eye, and miniature wing) that were obviously on the same chromosome as that which carried the sex factor. But before long Morgan found that all the dozens of mutations sorted or segregated into three groups, corresponding, he correctly supposed, to the three large chromosome pairs in *Drosophila*.

At one stage in 1914 there was some concern that only three linkage groups had been discovered, whereas there were clearly at least four visible pairs of chromosomes in *Drosophila*. True, the chromosomes of the fourth pair were very tiny, but nevertheless Morgan predicted the presence of a fourth linkage group and one of his students, H. J. Muller, soon found bent wing, the first gene on the fourth chromosome pair. Only a few genes were later found to sort or segregate with bent wing. Thus the number of genes in each linkage group was proportional to the length of the chromosome pair to which they belonged.

Probably Morgan held back the additional data on miniature wing because he had no explanation for the fact that while both miniature wing and white eye were "sex-limited," or sex-linked, and this was supposedly because these genes were both located on the sex factor or sex chromosome, the two characteristics sometimes sorted independently. In other words, the fly whose mother carried white eye and miniature wing on one chromosome produced some males with white eyes and miniature wings but also some with white eyes and normal wings and some with normal eyes and miniature wings. In fact, the two characteristics were inherited to some extent independently, despite both being presumably due to sex-linked genes.

Morgan then hit on an explanation for the failure of the two sex-limited genes to segregate together; perhaps they were not situated *close together* on the X chromosome. If during meiosis the two X chromosomes (like the other pairs of chromosomes) exchanged their genes, well-separated parts of the chromosome would be likely to exchange. For two genes close together, this was more unlikely.

The process of interchange of genes between chromosomes is called crossing over; the tendency of factors to stay together is linkage (both terms were coined by Morgan). These ideas may be easier to understand if the genes in a chromosome are considered as being like playing cards in a deck. Genes are distinct units like cards, which are "played" separately and do not blend. The chromosome pairs come to lie intimately together during meiosis, at which time the maternal and paternal chromosomes are shuffled so that their genes become exchanged in the same way as the cards of two decks of playing cards that are shuffled together. And the closer any two cards lie together in a deck the less likely that a single cut will separate them. Indeed there is only about a 2 percent chance of cutting between any two particular adjacent cards and a 98 percent chance of not doing so.

The possibility of this type of exchange of chromosomal material, which Morgan called crossing over, had been suggested by Sutton, Wilson, Lock, and Doncaster. But Morgan, in conjunction with his student A. H. Sturtevant, supplied genetic proof and added the original suggestion that the distance between genes (only a few millionths of an inch) and their order could be inferred from the results of genetic crosses. In other words, the greater the degree of independence of segregation, the further apart the genes could be assumed to lie on the same chromosomes. Complete independence of segregation, of course, meant the genes were on different chromosomes, or far apart on the same chromosome. Sturtevant wrote:

In the latter part of 1911 . . . I suddenly realized that the variations in strength of linkage, already attributed by Morgan to differences in the spatial separation of the genes, offered the

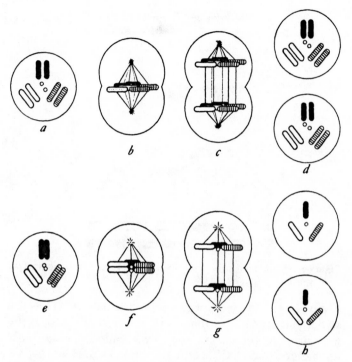

Figure 5. In the upper row of the diagram is a typical process of nuclear division such as takes place in the body cells or the early germ cells. In the lower row is the separation of the chromosomes that have paired. This sort of division takes place in meiosis, so that mature germ cells have only one of each pair of chromosomes. Reprinted from Morgan, *A Critique of the Theory of Evolution*, figure 49

possibility of determining sequences in the linear dimension of a chromosome. I went home and spent most of the night (to the neglect of my undergraduate homework) in producing the first chromosome map, which included the sex-linked genes y [yellow body color], w [white eye], v [vermilion eye], m [miniature wing], and r [rudimentary wing], in the order and approximately the relative spacing that they still appear on the standard maps.

Morgan and his group began to measure, or rather infer from the frequency with which genes crossed over, the

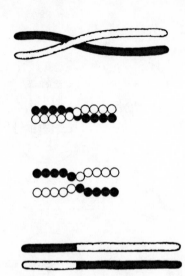

Figure 6. Simple crossover. Reprinted from Morgan, *A Critique of the Theory of Evolution*, figure 64

distance of all the mutant genes from each other. They did so with such accuracy that their chromosome map, as the gene array on the chromosomes was called, is essentially unchanged even half a century later. J. B. S. Haldane, the English geneticist, suggested that these units of measurement be called *morgans*. For example, Sturtevant found the gene for yellow body to be 1.5 centimorgans away from the gene for white eye, which in turn is 5.4 centimorgans away from that for bifid. Since bifid and yellow were 6.9 centimorgans apart, they had to be on opposite sides of white (Figure 7).

The theoretical description of the first work on linkage was published on September 10, 1911, in *Science* and under Morgan's name alone. It is generally considered to be one of the two most important manuscripts of the original *Drosophila* work. In it Morgan wrote:

Mendel's law of inheritance rests on the assumption of random segregation of the factors for unit characters. The typical proportions for two or more characters, such as 9:3:3:1, etc., that

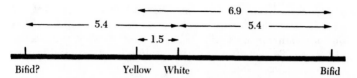

Figure 7. Plotting the gene array on a chromosome

characterize Mendelian inheritance, depend on an assumption of this kind. In recent years a number of cases have come to light in which when two or more characters are involved the proportions do not accord with Mendel's assumption of random segregation. The most notable cases of this sort are found in sex-limited inheritance in *Abraxas* and *Drosophila,* and in several breeds of poultry, in which a coupling between the factors for femaleness and one other factor must be assumed to take place, and in the case of peas where color and shape of pollen are involved. . . .

I venture to suggest a comparatively simple explanation based on results of inheritance of eye color, body color, wing mutations and the sex factor for femaleness in *Drosophila*. If the materials that represent these factors are contained in the chromosomes, and if those factors that "couple" be near together in a linear series, then when the parental pairs (in the heterozygote) conjugate like regions will stand opposed. There is good evidence to support the view that during the strepsinema stage homologous chromosomes twist around each other, but when the chromosomes separate (split) the split is in a single plane, as maintained by Janssens. In consequence, the original materials will, for short distances, be more likely to fall on the same side as the last, as on the opposite side. In consequence, we find coupling in certain characters, and little or no evidence at all of coupling in other characters; the difference depending on the linear distance apart of the chromosomal materials that represent the factors. Such an explanation will account for all of the many phenomena that I have observed and will explain equally, I think, the other cases so far described. The results are a simple mechanical result of the location of the materials in the chromosomes, and of the method of union of homologous chromosomes, and the proportions that result are not so much the expression of a numerical system as of the relative location of the factors in the chromosomes. *Instead of random segregation in Mendel's sense we find "associations of factors" that are located near together in the chromosomes. Cytology furnishes the mechanism that the experimental evidence demands* ("Random Segregation Versus Coupling in Mendelian Inheritance," *Science* 34:384).

Morgan's triumphant paper was rather more theoretical than might be expected from a man who detested theories, and its complete absence of data was surprising for a man dedicated to the quantitative experimental method. (Morgan published the supporting data a year later, and two years later, in 1913, Sturtevant published the original map.) But the paper's very brilliance dispels any criticism, even the complaint that he should have given the date and source of the article by Janssens to which he alludes. This kind of omission is noted from time to time throughout Morgan's life, but quite probably it was due to his general untidiness rather than to any attempt to slight another scientist.

Now that the fusion between cytology and genetics has been taken as fact for so long, it is hard to perceive how far apart the two lines of development were. In England, for example, genetics studies were done without any cytology. Bateson admitted in 1921, "Cytology here [in America] is such a commonplace that everyone is familiar with it. I wish it were so with us." It is not surprising that the fusion should have occurred at Columbia, since Morgan sat in the office next to Wilson's—the one scientist doing genetic studies, the other cytological studies— and since students such as Payne were given *Drosophila* projects by Morgan and chromosomal analysis of insects by Wilson (Payne incidentally proved the chromosomal determination of sex in seven species of insects in 1908 and 1909).

With one of the world's leading cytologists next door, friendly and supportive, Morgan had access to the best cytological opinion in the world. While keeping abreast with recent research to which he contributed in large measure, Wilson carefully cataloged all work relevant to development and cytology in his magnificent book *The Cell in Development and Heredity*, various editions of which appeared in 1896, 1900, and 1925. Even today it makes good reading and provides an excellent reference source; Muller's introduction to the 1956 reprint adds additional illumination.

In 1913, as soon as Morgan was convinced of the meaning of the experiments with chromosomes, he wrote *Sex and Heredity*. In 1915 Morgan and three young colleagues—Sturtevant, Bridges, and Muller—who had worked with him for most of the preceding five years produced Morgan's best-known book, *The Mechanism of Mendelian Heredity*. It summarized all the *Drosophila* studies and was what Curt Stern has called the fundamental textbook of the new genetics. This book was a first attempt to reconcile all of genetics to chromosome behavior, and the material in it is now so much taken for granted and so basic to the edifice of modern genetics that it is difficult to appreciate exactly what a huge leap in understanding it represented. It established Mendelian laws and their exceptions and explained them in terms of factors (that is, genes) which were physically part of the visible chromosome in a linear array and whose behavior in genetic studies correlated perfectly with chromosome behavior. Genes are paired; chromosomes are paired. Only one of each pair goes to any one offspring. Genes exist in linkage groups corresponding to the number and size of chromosomes.

The book was a masterly and accurate evaluation, and an audacious one, considering how little was known at the time of chromosome behavior in *Drosophila*. All the the predictions of chromosome behavior made between 1910 and 1920 were based on genetic evidence, for the *Drosophila* chromosomes themselves could be seen only through a glass darkly (for example, in 1914 Morgan thought there were five pairs [*Popular Science Monthly*,

January]). Support was provided by analogy with other organisms.

The Mechanism of Mendelian Heredity was steadily recognized as valid across most of the civilized world, and in America the honors rolled in.[5] Johns Hopkins University presented Morgan with an honorary LL.D. degree, which he always used in the books that followed, and the University of Kentucky presented him with an honorary Ph.D., the first of several he was to collect. He became a member of the National Academy of Sciences and later its president. He was made foreign member of the Royal Society, and in 1924 received the Darwin Medal and the Copley Medal in 1939. As a consequence of these awards he found it easy to obtain money from groups such as the Rockefeller Foundation and the Carnegie Institution.

The success of the *Drosophila* work also meant that Morgan was established as the twentieth-century Mendel and attracted scientists and visitors from all over the world prepared to kneel at the shrine of some scientific deity. What they found in the Fly Room at 613 Schermerhorn Hall, Columbia University, was as scientifically wonderful as they had suspected, hoped, or feared, and a unique reflection of the man the students called "the Boss."

6

THE FLY ROOM

Should you ask me how these discoveries are to be made . . . I would then say: By industry. . . . By the intelligent use of working hypotheses (by intelligence I mean a readiness to reject any such hypotheses unless critical evidence can be found for their support). By a search for favorable material. . . . And lastly, by not holding genetics congresses too often.

Thomas Hunt Morgan
*Presidential address to
the International Genetics
Congress*

THE NAME Thomas Hunt Morgan appeared likely to live in history, his father's wish. And when family regretted the absence of descendants bearing the name (since Morgan's only son's children were daughters), they recalled the remark of Morgan's step-grandson James Mountain: "Praise the name and pass the genes along." What is more important than either, however, is the cultural heritage that Morgan passed on to scores of young geneticists.

He had gathered around him a group of students remarkable for their intelligence and ability to work both individually and as part of the team. Morgan could have had his pick of the graduate students at Columbia University, many of whom did indeed have space at one time or another in the Fly Room or its more spacious environs. But he had an unerring sense of people, and no sense whatever of the snobbery and appeal of degrees. Morgan met Alfred Henry Sturtevant and Calvin B. Bridges in a general zoology class which he had temporarily taken over for another professor. Both were undergraduates and teenagers. Sturtevant brought Morgan a manuscript he had prepared on coat color among his father's and brother's race horses back on a farm in Alabama, and Morgan was impressed. He helped Sturtevant publish it as *Study of the Pedigrees of Blooded Trotters* and set him to work counting *Drosophila*. (Unfortunately, Sturtevant was color-blind, and so somewhat limited in his ability to add new color mutants to the growing number.) Within two years, when Sturtevant was only twenty-one, he had made an invaluable contribution by preparing the linear order of genes on the chromosomes, soon to be called maps.

In 1910 Morgan gave a young undergraduate, Calvin Bridges, a job as bottle washer in the laboratory. When Bridges spotted a vermilion-eyed mutant through the thick glass bottle, a color distinction most people have trouble making even with a microscope, he was immediately made Morgan's personal assistant, presumably paid out of Morgan's pocket. Bridges continued to spot mutants readily as well as some unusual inheritance patterns from which he inferred the failure of a chromosome pair to separate, which he called nondisjunction. Until he died in 1938, Bridges remained a close colleague of Morgan. Both Bridges and Sturtevant completed their bachelor's degrees and began doctorates under Morgan's direction. For seventeen years their chief occupation too was "to count flies for Columbia University."

Perhaps the best-known of Morgan's Fly Room students was H. J. Muller, who had already received his bachelor's degree from Columbia in 1910 and was working on a master's degree. During 1911 and 1912 he went to Cornell to medical school, but was back at Columbia to complete a Ph.D. degree and act as an assistant or instructor from 1912 to 1915 and from 1918 to 1920, with a stint at Rice University in between, initially under Julian Huxley. Although Muller did not share completely in the bond that held Sturtevant and Bridges to their old teacher, he was coauthor of the *Mechanism,* made many important contributions to the understanding of gene interactions, and went on to demonstrate that X-rays would increase the mutation rate 150 times—a discovery for which he was to receive a Nobel Prize in 1946. The fourth member, a kind of invisible partner in this collaboration of mind and talent and work, was Wilson. At the Sixth International Congress of Genetics in 1932 Wilson reminded the assembled delegates:

I am not and never was a geneticist and can be accepted as such only by courtesy. It may therefore surprise you if for a moment I boast of a certain achievement in that field, one for which I have never received any credit. More than forty years ago I discovered a new and strongly dominant Mendelian character who I was able to recognize as such long before the resuscitation of Mendel's work in 1900. That character is well-known to you all—it is at this moment the honored president of this genetics congress, Thomas Hunt Morgan.

Among all the workers in the Fly Room there seemed to exist a respect and ease not often found in a classroom or laboratory situation. Dr. Tove Mohr, the wife of O. L. Mohr, the first European postdoctoral student, was astonished by her first sight of the young Sturtevant lounging in a chair, a pipe in his mouth, his feet up on the desk, arguing with Morgan.[1] Sturtevant, the most articulate and admiring of Morgan's students, described the laboratory situation thus: "This group worked as a unit. Each carried out his own experiments but each knew exactly what the others were doing, and each new result was freely discussed. There was little attention paid to priority or to the source of new ideas or new interpretations." No one who ever studied under Morgan has failed to mention the great pains he took to explain things to young students— and the ease and camaraderie he felt with them. Dr.

Sewall Wright's story of being hoisted over the locked door of a men's room stall by Morgan is but one of many illustrations of this friendly feeling.

The small Fly Room had eight desks crowded into it. There was also a kitchen table where the janitor and later a trusted student prepared the *Drosophila* culture medium. At first this was mashed banana, fermented the way the flies preferred it. The resulting smell was fierce and drew constant complaints from the rest of the biology department. Later Morgan discovered that banana juice was cheaper than whole bananas, and yet later commercially prepared food proved most efficient and economical. Agar was also used. A rotatable pillar bore chromosome maps on its four faces, and notes on the various types of chromosome rearrangement were penciled in.

Near the entrance of the room a stalk of bananas hung conspicuously, serving as a center of attraction for the numerous fruit flies that had escaped from their milk bottles or that had bred themselves, without the benefit and direction of science, in the garbage can that was never thoroughly cleaned. The bananas were sacrosanct except to Wilson, who alone had the privilege of eating one. (One of Morgan's children says Wilson was repaid, however, when Morgan and a group of students once made an omelet out of an ostrich egg Wilson had obtained especially to be photographed for the cover of one of his books.)

The room also had other wildlife. The agar for the *Drosophila* medium was the permanent residence of hordes of cockroaches. When the drawer was opened the agar moved. Curt Stern, who worked with Morgan in the 1920s, wrote, "During the years I worked in the inspiring spiritual atmosphere of the Fly Room I never opened my desk drawer without looking away for a while to give the cockroaches a chance to run into the darkness. Once I said breathlessly, 'Dr. Morgan, if you put your foot down you'll kill a mouse.' He did!" Stern referred to the haphazard and make-do simplicity of the Fly Room as "little science" and added, "Some years later, Morgan founded a well-equipped modern laboratory at the California Institute of Technology, but during the Columbia years little science was the appropriate way to make scientific progress instead of assigning time to technical perfection."

It was also cheap. Morgan's stinginess with institution funds—though matched by his generosity with his own—was legendary. Not only were many of the *Drosophila* containers makeshift, so was much of the equipment. When equipment entered the laboratory, it was frequently over the Boss's objections. Hand-lenses were only gradually replaced by simple microscopes, and the microscope electric light shields were constructed out of tin cans. When the roof leaked, buckets were placed on the floor, and when the flies began to suffer from the cold, the inventive Calvin Bridges was permitted to make a simple thermostat. When his experiments made *Drosophila* famous, requests for stock poured in and Morgan sent cultures out freely—and free. But he expected this unusual generosity with Columbia's funds to be reciprocated. He wrote Cole of Wisconsin to ask for a stock of pigeons: "Cole: I am not offering to pay for the pigeons or the freight. We have honored all requests for *Drosophila* from all over the world without so much as a postage stamp. So send me the pigeons. Morgan."

The flies (also cheap; a dime a day fed the whole crew the first year) were threatened with a dramatic end the evening the gymnasium next to Schermerhorn Hall burned. Morgan rushed out into the winter night where horsedrawn fire engine pumpers were playing water on Schermerhorn to protect it from the blaze. Some of the windows had already melted from the heat and the *Drosophila* stocks were in that immediate area. Police and firelines were holding spectators back, but Morgan talked his way past the police and raced up the six flights to the Fly Room on the top floor. He could never have brought down all those small containers, but he removed them from one end of the building, which had already grown suffocatingly hot, to the end furthest from the fire. Then he reluctantly left them to watch from the sidewalk until the fire at the gymnasium was out. Luckily the fire did not spread; the flies were safe.

In the middle of the confusion and chaos and filth that typified the Fly Room, the exacting work continued calmly. Dr. Morgan counted his flies, using a jeweler's loupe, as he stood behind a desk cluttered with opened mail. When the stack became too high, he usually pushed it onto an adjacent desk of one of the students who, once the Boss had left the room, pushed it back, a seesaw that continued until someone made the decision for Morgan and threw the whole lot, sometimes unanswered, into the garbage.

Morgan's desk was even more unseemly than this would indicate. Most of his co-workers threw discarded flies into a jar of oil called the morgue. Morgan merely squashed them on the porcelain counting plate that was already covered with the mold of the week's previous counts. Occasionally the wife of one of the graduate students (wives often held jobs tending the stocks as a kind of extra fellowship) would timidly wash the half-dried flies off the great man's plate, but he was suspicious of the gleaming porcelain and squashed the next day's flies on it with vehemence.

There is little doubt that Morgan enjoyed exactly this kind of laboratory. He was by nature a careless and sloppy man, but he also delighted in shocking others, in playing the imp. It was this mixture of eccentricity and disregard for opinion that made him sometimes tie up his pants with a string when he couldn't find his belt, or show up without any buttons on his jacket. Once when he found himself with an obvious hole in his shirt he asked someone in the office to paste over it with a little piece of white paper. More than once Morgan was mistaken for the janitor, but he always managed an air of elegance even at his most purposeful dishevelment, which could be considerable.

Perhaps the first edition of *The Mechanism of Mendelian Heredity* (1915), in which the Morgan group established the chromosome basis of inheritance and proved that genes were part of the chromosome in linear array, summed up the most important contributions to genetics that Morgan would make personally. But over the next ten years he continued as the center and the catalyst of numerous research projects at Columbia.

A whole host of discoveries followed. Morgan and his colleagues came to realize that a distorted sex ratio could be explained as the result of a sex-linked lethal gene. Instead of half the males receiving a gene that gave them white eyes, they received a gene that was lethal, so that one normal male survived for every two females. The team also found that while one crossover might occur at any point at random on the chromosome, a second crossover was unlikely to occur near the first. This phenomenon they called interference. In addition they discovered that crossing over does not occur at all in the male *Drosophila*. Bridges showed that the frequency of crossing over changes with maternal age. His discovery of nondisjunction (the failure of a pair of chromosomes to disjoin or separate, causing the daughter cells to receive an abnormal chromosome complement) was one of the most dramatic of the group's early findings—and very convincing to any who still doubted the chromosome hypothesis. The work on nondisjunction led to discovery of *Drosophila* with different numbers of chromosomes, such as seven, nine, or ten. While the XX-XY chromosome explanation of sex determination was adequate for flies with eight chromosomes, it required modification for those with a variant chromosome number. Bridges then proposed his balance theory of sex.[2] Under this hypothesis, the Y chromosome was not male determining. Given two pairs of autosomes (chromosomes *other* than sex chromosomes), one X, with or without a Y, gives a male; two Xs, with or without a Y, a female; and 3 Xs, with or without a Y, a superfemale. Given three pairs of autosomes, one X gives a supermale, 2 Xs an intersex, 3 Xs a female, and 4 Xs a superfemale.

Our present view of sex, which has in the main derived from the Columbia work, has established that in organisms having paired chromosomes males are the heterogametic sex (XY) in Mammalia, Nematoda, Mollusca, Echinodermata, most Arthropoda, and dioecious plants. In Lepidoptera and moths, most Reptilia, primitive Trichoptera, some Amphibia, some fishes, some Copepoda, and probably all birds, the female is the heterogametic sex. Most fish are without a chromosomal sex-determining mechanism, and in some species it is still not certain how sex is determined. In fact, the role of the sex chromosomes in determining sex in humans—XY for males, XX for females—was not established with certainty until 1960, when it was found that a single X with no Y resulted in an infertile female and a Y with no X was lethal.

Sturtevant also made several important contributions, of which mapping was only the first. He came up with the explanation of multiple alleles. He also inferred the existence of inversions, in which a bit of chromosome breaks off and is stuck back upside down. Since this was fifteen years before the giant banded salivary gland chromosomes proved the prediction true, the prediction itself could be listed alongside John Couch Adams's prediction that Neptune must exist, made three years before anyone ever saw it.

Morgan's students remember him as a happy man. He was doing exactly what he most enjoyed, whether it was the world-famous *Drosophila* work or some experiment whose purpose no one ever figured out exactly, like the one in which a crab walked around with another crab glued to its back, a fragment of radium between the pair.

One of the topics to which Morgan kept returning during this period, even after his dramatic conversion to Mendelism, was the possibility of environmental determinants in heredity. Occasionally he would publish an article showing some curious violation of Mendelian law in adverse conditions, and one could not help feeling that the chromosome theory still reminded him of the preformation theory, which he was pleased to disprove. For example, the condition abnormal abdomen was influenced by the degree of moisture in the food, and both leg reduplication and vestigial wing were influenced by temperature. He also continued his embryological studies, particularly investigating self-sterility in *Ciona* and the fiddler crab's asymmetrical claw. He also studied the inheritance of the number of tail feathers in fantail pigeons that Cole sent him and the mechanics of the production of secondary sex characteristics in chickens and fiddler crabs.

Morgan continued to teach graduate courses at Columbia, although his attitude toward teaching had not changed. "Excuse my big yawn," he once apologized, returning to the laboratory, "but I just came from one of my own lectures." He felt the same way about administrative duties. When L. C. Dunn was debating whether or not to come to Columbia, he told Morgan he feared the atmosphere was wrong there, and Morgan summed up his own attitude. It was true, he said, that there were certain biological requirements to teaching at a large university. "For one thing, you have to reverse your evolution and develop an exoskeleton, and you have to learn to keep out of corridors because corridors generally lead to committee rooms." He then asked Dunn if there was more than one chair in his office and shook his head sadly on hearing there were two. "That's a mistake. There should be only one, and you should sit in it."

And sit Morgan did. For a decade and a half he worked almost unrelentingly in the Fly Room. And so did Bridges and Sturtevant, who made up with Morgan the nucleus of the changing Fly Room group.

Morgan, naturally, was very eager that the group stay together, although Sturtevant always emphasized,

"Morgan did not direct our work. This was characteristic; he did his own work, and had no desire to develop a group working under his direction." But the group's success and, quite important to Morgan, its spirit of play and mutual endeavor, were unique. After 1915 Morgan received a grant from the Carnegie Institution to support the Drosophila work (a grant that would continue until his death), part of which he used to support Bridges, Sturtevant, and later Jack Schultz as full-time research assistants. Not only did they accompany the Morgans to Woods Hole each summer, but when Morgan took a year's leave from Columbia to work at Stanford University, along went Bridges and Sturtevant, a host of graduate students—and also Miss Edith Wallace who did the lovely and delicate drawings of the variant flies. And when Morgan left Columbia, there was no question what would happen to the two men.

H. J. Muller, the fourth coauthor of the *Mechanism*, had left for Texas in 1920, where he sometimes openly criticized Morgan for not giving his students and indeed his co-workers full credit for their individual discoveries, allowing them to seem the work of the group. It was not always Morgan who convinced his students and then the world of the brilliant conclusions of the Fly Room, according to Muller, but sometimes the students who convinced the skeptical and occasionally uncomprehending Morgan—who then was better able to convince the world than any of them might have been.

Morgan was hurt, but typically he feigned calm and indifference. He wrote a colleague in 1934: "Muller's attitude has always been antagonistic to us, although he has generally managed to keep this under cover and we have consistently ignored it, treating him in the most friendly way, because we regarded his attitude as wrong and inexcusable." The Mohrs, friends of both, considered it wrong yet understandable.

To Bridges and Sturtevant, Morgan acted almost as a

father. The boyish, likable Bridges, in particular, sometimes seemed in need of one and must have stretched even Morgan's unusual degree of tolerance. The younger man was an advocate of free love and practiced what he preached. But Morgan, who would bear no discussion of a man's politics, religion, or private life, held to this rule with Bridges. Himself almost puritanical in matters of sex, he nevertheless defended Bridges and his dramatic (and by the standards of the time scandalous) behavior as having no bearing on his place in the laboratory.

Morgan skillfully held his team together, continued innumerable experiments, received an increasing number of geneticists from all over the world, forged plans for uniting physics, chemistry, genetics, botany, and zoology, and continued to turn out scores of articles and books. He wrote continually, and as soon as he got results he published them. In 1916 he published *A Critique of the Theory of Evolution,* based on the Louis Clark Vanuxem Foundation Lectures he gave at Princeton, and in 1919 he included his book *The Physical Basis of Heredity* in the Monographs on Experimental Biology series he and Loeb were editing for Lippincott. Both of these books were revisions of earlier material. Still other books followed, including *Evolution and Genetics* in 1925 and *The Theory of the Gene* in 1926. *Experimental Embryology,* published in 1927, was 750 pages long, an exhaustive and painstaking review of the formative stages of development—and unusual for Morgan in that it was carefully written and researched. The bibliography alone was 100 pages long. Strangely enough, despite its excellence, the book received little notice then and remains virtually unknown. This must have puzzled Morgan, who had been criticized for the sometimes haphazard construction of other books.

But if Morgan was not being given his due as an embryologist, he was given full glory as a geneticist. The 1915 *Mechanism* had been the introduction to genetics for many a young scientist; *The Genetics of Drosophila,* written with Bridges and Sturtevant and published in 1925, proved to be the Bible for those who had grown up to be geneticists. Morgan's name (and for that matter, his fortune) were made, and his successes inspired a continuing chorus of praise. "Morgan's theory of the chromosome represents a great leap of imagination comparable to Galileo or Newton," wrote C. H. Waddington, for example. C. D. Darlington credited Morgan with first joining "the gulf which has always sundered body from egg and likewise mind from matter by a bridge of experimental method. . . . the chromosome theory begins to appear as one of the great miracles in the history of human achievement." H. J. Muller wrote that "Morgan's evidence for crossing over and his suggestion that genes further apart cross over more frequently was a thunderclap: hardly second to the discovery of Mendelism, which ushered in that storm that has given nourishment to all our modern genetics." And Morgan himself had a favorable view of the work: "Since 1900, when the discovery of Mendel's work became known, one of the most amazing developments in the whole history of biology has taken place."

7

AT THE MORGANS'

*On being asked by Punnett what he would like to see in
England, Morgan replied, "I want to hear the skylark
which we haven't got in the States. . . . Let us talk,
and then the skylark."*

Dr. Tove Mohr

THE FIRST FIFTEEN years at Columbia were the most
fruitful of Morgan's professional career, and the most
obsessive. He had always been an extremely hard-
working man, with a clear perception of priorities, but
once he saw the white-eyed fly his life calcified into even
more unrelenting patterns. His first and best, and sometimes last, energies went to his work.

And yet this same period encompassed the first years of
his marriage, the birth and childhood of four lively children with as vast a collection of pets as could be expected
from the offspring of two naturalists, the buying of one
house and the building of another, and the accumulation
of the quantities of paraphernalia necessary for the almost
Victorian life still lived in turn-of-the-century New
York.

That he was able to manage his personal affairs so well
was in many ways a tribute to the woman he married. She
saw to it that he didn't have to.

At the beginning, when the Morgans moved into a
rented house only a five-minute walk from Morgan's

office and laboratory in Schermerhorn Hall, Tom and Lilian were together constantly. Lilian accompanied him on his field trips—"tagging along after Tom again," she wrote his mother—and tried to spend some time in the laboratory herself, working to the side, staying out of everyone's way. In the evenings at home they sat at opposite ends of the table and wrote. Those first years— before the flies—were also heavy with friends and visiting family. Lilian managed to persuade her husband to attend concerts, which she loved, and he taught her to ice skate. They attended a rash of dinners and parties, with Lilian's family in New York and with their "biological" friends from the university or the Museum of Natural History.

A year after their marriage, Lilian was pregnant. Tom's mother rushed his own little rosewood cradle northward and began stitching by hand delicate baby garments just like the ones she had made for Tom himself. Howard Key Morgan was born February 22, 1906, in New York City and was surely one of the most adored and best-dressed infants ever known. "You asked how Tom was with the baby," Lilian wrote her mother-in-law. "I can only whisper to you that he is as ideal as a father as he always has been as son and husband." He tended to excite the babe so much, in fact, that the reluctant decision was made to keep father out of the nursery before bedtime. In the summer Howard was carried off to Woods Hole, to rented rooms in a lobster fisherman's house.

Even with the new baby, Lilian was able to spend some time in the laboratory. "There are several persons here working under Tom's guidance," she wrote his mother, "I being one!" But it was difficult for her to get her strength back. She was thirty-six when the baby was born, and there was now another confusing and exciting element in their lives. They were designing a large house to be built at Woods Hole on Buzzard's Bay Avenue on Sandy Crow Hill, then a large potato patch about three-quarters of a mile from the laboratory. It was a house built for work—and for a large family. Besides the enormous living room, part of which could serve as a guest room, there also were six bedrooms, two sleeping porches, and a third floor with at least four rooms for the maids who were, in Woods Hole as in New York, an important and necessary part of the Morgans' life. Not only would Tom's family spend time each summer at Woods Hole, but his students and friends from New York were constantly in and out, as well as distinguished guests and visitors. Sometimes the house would have seventeen people sleeping in it and waiting cheerfully in line for the one bathroom.

The following summer the family returned to Woods Hole and the newly completed house—with a few students and a maid and a nurse for Howard. Lilian, however, settled in early with her own nurse at an inn in New Bedford, across the bay, and waited for a second child to be born. Morgan traveled across by boat to visit Lilian and to bring news of the experiments at Woods Hole. On June 25, 1907, a girl was born, to be named Edith Sampson Morgan, after Lilian's much-loved sister.

The enlarged family came back to New York to the new house they had bought down the street from their rented one (so as not to lose the convenience of location). They were still only a few minutes from Schermerhorn Hall and three blocks away from the Wilsons. After Morgan left Columbia, the house at 409 West 117th Street was bought by Columbia and turned into the Institute of the Study of Human Variation, and experiments and research were carried on where Morgan used to lean against his junior-size billiard table.

A third child, Lilian Vaughan, was born in New York on January 5, 1910. The fourth and last child was born August 20, 1911, in New Bedford, while Morgan and the first three—and to be sure a good staff of household help—visited Woods Hole. This child was named Isabel Merrick, after Lilian's mother Isabella Merrick. (Isabella's uncle Samuel Vaughan Merrick had been the first president of the Pennsylvania Railroad.) The Morgans obviously liked traditional family names. It could have been different. When Drs. Otto and Tove Mohr had a baby girl Morgan wrote: "Congratulations. I have only one piece of fatherly advice to give you. Don't name her *Drosophila*. I have resisted the temptation three times."

With a household of four children, and with Morgan himself entering a period of new and exciting work, the life of the Morgans changed greatly. Morgan stopped much of the traveling he had done earlier. "As you can see," he grumbled, "I can't very well travel with this raft of kiddies." But the kiddies were mainly just an excuse, since Lilian would have done the packing and probably would have liked it. They also curtailed much of their social life. Morgan had a standard reply to any invitation: "Take the children where you will," he told Lilian, "but I must go to the laboratory."

Lilian continued to see friends, particularly those with whom she might play the violin. She enjoyed the weekly science sessions for students and colleagues held at their house, and she quietly followed her own cultural and political interests. Some of these were far more liberal than her husband would have supported, or perhaps even approved, had she not been careful to spare him knowledge of her activities. She believed fervently in the League of Nations. At heart she was an ardent suffragette—but saw no need to trouble Morgan with this information. Her involvement with causes was more often financial than physical. Money came easier than time, and time went to the support of her husband's complex working life.

The household centered on Morgan's working schedule, in fact. Seven days a week he rose late and breakfasted alone after the flurry of children. He walked to the laboratory. At noon he came home for lunch with the family, and then returned to work. At five o'clock sharp he played handball with a group of colleagues who met at the Columbia gymnasium. The age and academic accomplishments of his loosely organized team often fooled new students who took them on. Morgan approved of exercise and took it with a passion. After an hour, he returned home, often with cuts and bruises to show, and had a glass of whiskey before the family dinner hour. During Prohibition he made his own wine out of dried figs.

After dinner he played with the children. When they were little, he played on his hands and knees; when they grew to the questioning age, he answered them patiently—unless it was something they could look up themselves. He told marvelous bedtime stories, detective and cowboy thrillers to Howard, and fairy tales to Edith, Lilian, and Isabel. He often sketched out little illustrations to go with his stories.

As the children grew older, he let them play their own games but nevertheless kept them company, his mind split between the gamboling on the floor and the copy of *Science* or some other journal on his lap. And when the children were in bed, he retired to his study, which had been added on the roof or fifth level of the large house. Here he wrote until late into the night, while Lilian kept *him* company. She sat on the study couch reading or mending or writing letters. She never interrupted his work. They simply enjoyed being together.

On Sundays the same schedule applied except for the handball, so that Morgan came home from the laboratory early, making it a special day for the children. The family did not go to church. Both Tom and Lilian were indifferent ("beautifully indifferent," one of his children has said) to religion. Morgan was militantly opposed to religion, however, when it got in the way of knowledge, as it did in opposing evolution and his own mechanistic interpretation of cellular behavior. But religion as expressed in the Episcopal church in which both he and Lilian had been baptized was something that he felt could be dismissed as one of his mother's and sister's fanciful quirks of character, in the same category as their reverence for Civil War heroes and the Morgan name. He would tease them as they set off faithfully and unfailingly to church at Woods Hole, but he nonetheless allowed his children to be baptized at birth, delaying the ceremonies until his mother could come north to witness them. And Lilian, who had painfully lost her faith while in college, dutifully sent her mother-in-law Easter cards crowded with cherubs. If the children asked to go to Sunday school, they were allowed to go. But they obviously did not go often. When Jerusalem fell to the Allies in 1917, the *New York Times* headlined the expulsion of the infidel from Gethsemane and Isabel, the youngest, had to ask her father what Gethsemane meant. He smiled at her and lovingly said, "Oh, you little heathen."

During this period, Morgan obviously loved and depended on his family, but his strength as a father was in special events. He might be silent at the supper table, thinking ahead to the evening hours and the book or article or report to be written. But he was unforgettable at Christmas. The whole family walked down to Amsterdam Avenue to buy just the right tree and decorated it with delicate balls and tiny candles. Then the children waited breathlessly for the appearance of Santa Claus, dressed in an old red bathrobe trimmed in fur, with a pillow-shaped belly and a long flowing white beard. Santa Claus, the children always remembered, had extremely bright blue eyes. He distributed presents generously from a bag over his shoulder. (Some had a moral: scissors for all the children, with a note to cut out the squabbling.) The children always regretted that their father never got home from the laboratory in time to meet this marvelous figure. Years later, when grandchildren began arriving, the bathrobe and beard reappeared from some trunk and Santa's blue eyes sparkled again.

But the main event of the young Morgan family life during these years—and the one time when the whole family became part of the laboratory life for a brief moment—was the transfer each year of Thomas Hunt Morgan from New York City to Woods Hole. The trip took place the week classes ended at Columbia, but it involved planning for weeks, and planning meant Lilian. She must have longed for the first year she and Tom were married, when during the last week of school they camped out in the study of their New York house, eating their meals at a restaurant across the street and letting the dust pile up in the unused rooms while they read and prepared for the research to be done in the summer ahead, he at one end of the laboratory table, she at the other.

Now there were from six to eight trunks spread open upstairs, all carefully packed with petticoats and diapers. One monster trunk was filled completely with Tom's books, a horror for the men who had to carry it from the fifth-floor study to the horse-pulled carriage on the street (and, come September, back up again). Another trunk was left half-empty, since Morgan would invariably think of half a trunk's worth of last-minute packing.

But most of his packing time was spent at the laboratory, lovingly and carefully organizing the *Drosophila* into their traveling vials with a little mashed banana in the bottom to sustain them for the trip.

The morning after school ended, the carriage pulled up for the luggage, and the family walked toward the subway. It was a confusion of children and servants, of plants that had to be carried, of goldfish and parakeets, of a disobedient and excited English setter. From the laboratory came the students and co-workers and a complex arrangement of chicken, mice, rats, all in steel cages, and the flies in their vials. As soon as the children were old enough, they were given *Drosophila* to carry. Some members of each fly stock always remained in New York on the trip out and in Woods Hole on the trip back. The first thing Morgan did on arriving safely was to telegraph an assistant that the stock was safely relocated. Only then could the home flies be moved or allowed to perish without continuing the family line.

On the overnight boat from the Hudson River pier to the Cape, whence a train carried the Morgans on to Woods Hole, the family ate from a large, carefully prepared dinner basket and tried to heed Lilian's admonishments to calm down. Morgan spent most of his time on deck to watch for storms, and to check constantly on the safety of his creatures.

The schedule at Woods Hole was a somewhat lighter replica of the one in New York. After breakfast, Morgan pumped up the tires of his bicycle and was off to the lab. At noon he returned, and everyone at the house—visitors as well as family—went for a mandatory swim. The only exceptions, even later in life, were Morgan and his mother and sister. Lunch was the gathering time. Everyone sat on the huge veranda that enclosed three sides of the house or on a great fifteen-foot-wide staircase leading toward the meadow at the back. It was an hour of rapport but not of gossip or personal chitchat. At the end of the hour Morgan returned to the laboratory. He came home for supper, romped with the children a bit, then read and wrote far into the night.

Both of Morgan's parents, as well as his sister, came to the large open house at Woods Hole each summer, although by the time the house was built Charlton Sr. had found his first secure job (with the United States Revenue Service) and thus had only a brief vacation to spend. Nevertheless, in these two weeks he and his son seemed to enjoy each other more than before. Charlton was proud of the way people treated Morgan, and Morgan made it a point to introduce his father to some of the southerners at Woods Hole and especially to a man who knew of Charlton's work as consul. The older children remember liking their gentle grandfather, but the youngest, Isabel, was only an infant when Charlton fell ill in Lexington. After four months he died, on October 10, 1912, with his wife and daughter by his side. Confederate veterans turned out en masse for the funeral services the following day when Charlton Hunt Morgan was buried with the military honors he so prized.

After Charlton's death, Morgan's mother and sister became even more involved with the family. Lilian was extremely gentle and protective toward the two women. To the Morgan children, who lived a close and protected life in a family where emotions were not easily expressed nor personal remarks exchanged lightly, Grandmother Morgan and Aunt Nellie were a constant surprise and joy. Both were frail women, very charming, and with the same blue eyes as Morgan himself. But they showed fewer of the restraints and less of the self-discipline that governed both Tom and Lilian. The two Nellies were always full of special surprises, such as great gift boxes that opened into each other, each layer elaborately wrapped, with wonderful presents inside. They brought trunks filled with lovely dresses into a family where clothes were not especially important. They also brought wonderful stories, family legends, civil war romances, personal accounts of slaves moving through Lexington on the underground railroad, ancestors clashing swords and writing songs, and of course John Hunt Morgan's famous escape from prison.

After their father's death, Morgan and his younger brother Charlton had a somewhat improved relationship. Charlton once made the long trip to Woods Hole, charming the children to whom he had been a mysterious figure, seldom mentioned. He then owned and operated a laundry in Birmingham, Alabama. His fiancée Mary Tinklepaugh had a sick mother, with whom she remained— in much the same way as Nellie and many another well-bred young southern girl remained with a widowed parent—so that she and Charlton were engaged for twenty years. During that time Charlton made a will that left all his money to his fiancée, with the stipulation she not marry. When her mother died and they married they forgot about the will. Charlton died in March 1935, soon after his wedding, leaving Mary a widow and thus unable to inherit. The Morgans all gladly signed over their shares of his estate to Mary, but it was a legal tangle.

In August of 1924, Morgan's mother became ill while visiting the family at Woods Hole, and in fact remained there, too ill to travel, after the rest had to return to New York. She later returned to Lexington, with Nellie hovering anxiously by her side, and died the following January 15. She rated the local paper's main editorial, entitled "A Gentlewoman," which mentioned her beauty and her loyalties: "During all her life in Lexington Mrs. Morgan was known as the friend and comforter of the Confederate soldiers and of the families of Confederate soldiers. A devoted member of her church, Christ Church Cathedral, she was almost if not quite as devoted to the cause for which her husband fought and her family suffered. She was one of the first presidents of the United Daughters of the Confederacy." The obituary also referred to her distinguished son.

Tom and Lilian went to Lexington for the funeral. It was the first time some of the Kentucky cousins had ever seen Thomas Hunt himself.

Nellie Jr. continued to live at the house on Broadway, which she and her mother had divided into apartments after Charlton Sr.'s death. Morgan had long paid the taxes on the house as well as sent money regularly and helped cover the doctor bills both women incurred. Miss Nellie was a strict landlady, although most of the tenants, themselves older women, became longtime friends. Once she threw out a younger tenant, Dr. Caroline Scott, who had given a particularly successful party, on the grounds that Dr. Scott's mother and aunt would certainly *not* have approved. (The guest list was distinguished, the party over at 11:30 P.M.) Miss Nellie remained a prominent figure in Lexington, particularly among the members of Christ Church, and was a loving aunt to her nephew and nieces and later to their children.

Nellie outlived her famous brother. She died on January 24, 1956, after a long illness, in a Lexington nursing home. She was eighty-three. In the brief notice of her passing there was no mention of either Thomas Hunt Morgan or of her uncle John Hunt Morgan. But on the same page was a quarter-page advertisement of Kentucky whiskey, embellished with a sketch of the Thunderbolt of the Confederacy in the expectation that his exploits more than ninety years earlier would now sell whiskey.

When there had been only one child and even when there were two, Lilian Morgan had managed to work a few hours in the laboratory by remaining on a close schedule and relying on the three or four servants of the household. But by the time there were four children, the last born in her forties, she retired from the lab, although she kept up with her scientific reading and surveyed daily what was happening with the *Drosophila* work. Much of her energy went into protecting Morgan from excessive demands from the family and from life itself. She also took most of the responsibility for the children's education. She taught the girls to sew and mend and she taught Howard carpentry. Even as a girl she had designed various pieces of furniture for others to build; now she and Howard built a few pieces themselves. She kept the two youngest girls, Lilian and Isabel, at home until they were ready to begin third grade together at the ages of seven and nine, mainly because she thought it dreadful to confine children indoors during the middle of the day in winter. The children all attended private schools. Lilian encouraged all the children to study music, even though the general family consensus was that there was no musical talent in the family, their mother included. Nonetheless, she continued to play the violin weekly all through her life, often with a friend accompanying her on the piano.

Life was always comfortable at the Morgans', as it should have been. Morgan was not kin to John Pierpont Morgan for nothing. In addition to his salary, royalties from his books, and hefty fees for lectures, Morgan and his wife both had income from stocks and bonds. But certainly Morgan did not flaunt his comfortable financial position. He sometimes dressed like a ragamuffin and even his oft-cited personal generosity was hidden from view and usually anonymous. And while the family lived well, with excellent food and a plenitude of household help, it also practiced certain spartan economies. One of the children recalls, for example, that they bought their Christmas trees on Christmas Eve because that was when the trees were reduced.

In 1920, Morgan took his first sabbatical leave. He went to Stanford University's Hopkins Marine Laboratory in Pacific Grove for the summer and to Stanford University for the academic year. The next summer he spent at the University of California at Berkeley. It was the first time since their honeymoon at Pacific Grove seventeen years earlier that Tom and Lilian had gone anywhere together that far away and for that long a time, and for the children—raised only between New York and Woods Hole without even a trip to distant Kentucky—the experience was eye-opening. It was the first time any of them acquired any understanding of the facts of life, which they garnered not from their naturalist parents (Lilian's attempt at explanation being extremely theoretical) but from the open-minded children of the California sunshine. Furthermore, the family's rigid schedules were disrupted and new adventures allowed to happen. The Morgans bought their first car, an Overland, when they reached Pacific Grove, and Howard and his mother learned to drive it from the instruction manual. Howard learned quickly and when he was fourteen became the official driver. Morgan himself had no time or interest for such things. Besides, he knew he could depend on his wife or one of his students or even one of his colleagues to drive him anywhere he wanted to go.

The car meant an undreamed-of freedom for Lilian and

Lilian Vaughan Sampson in her wedding dress, 1904 *(Courtesy of the Morgan family)*

Passport photograph of the Morgans taken for their trip to Stockholm in 1934   *(Courtesy of the Morgan family)*

Thomas Hunt Morgan as a twenty-three-year-old graduate student at Johns Hopkins University in 1889 *(Courtesy of The Johns Hopkins University)*

Thomas Hunt Morgan, Ph.D. This photograph was in the 1891 yearbook at Hopkins.

Morgan house at Woods Hole, Massachusetts, about 1918 *(Courtesy of Dr. Tove Mohr)*

Bridges (foreground) and Sturtevant pitching horseshoes at Woods Hole *(Courtesy of Dr. Tove Mohr)*

Morgan with neighborhood children the day his Nobel Prize was announced (*Associated Press photograph*)

Edith, Lilian, and Isabel Morgan in Morningside Heights *(Courtesy of the Morgan family)*

Howard Key Morgan at about fourteen with the Morgan dog Kai
(Courtesy of the Morgan family)

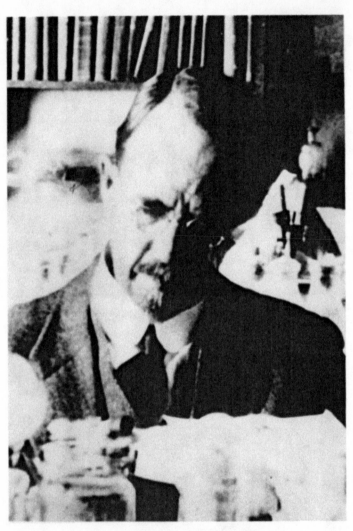

Morgan in the Fly Room. "Who combed my hair?" Morgan asked when he saw this snapshot.   *(Courtesy of Dr. Tove Mohr)*

the children. They used this year of rented houses and changed schedules to explore California on a series of camping trips. Their father went camping only once, on an elaborate all-male expedition. His excuse was always his work, but his children had begun to suspect that part of the reason was also his fondness for his own routines and the comforts of home, for the cigar after dinner in a rocking chair (in which no one else would have dreamed of sitting), the journal at mid-evening, and the papers in the lamplight late at night. And also the company of his students and colleagues at Columbia, many of whom had accompanied Morgan on his cross-country trek.

Not long after their return to New York in 1921, and before they had fully settled back into the old patterns, Lilian decided that her family no longer demanded so much of her attention. All of the children were in school, and the household was running well. She returned to the laboratory for about five hours a day. In the mornings now, when the children set off to school, Lilian walked with her husband toward Schermerhorn Hall, both of them returning to lunch with the two youngest children.

It must have been somewhat awkward for Lilian. It was awkward enough for any of the women graduate students in biology at Columbia. They did not work in the inner sanctum of the Fly Room but in an outer room, a kind of second circle. Female graduate students were not invited to the Morgans' Friday night biology readings, although this does not seem to have dampened their admiration for Morgan. But Lilian was in the even more awkward position of having no real place in the laboratory. She was not a student. She was not her husband's assistant, as is sometimes said, nor was she in any real sense his associate. Her work was unpaid and her use of the stocks and equipment only an unspoken agreement. Her research topics were separate, although closely related to his or to those of some member of the group.

In 1921 she was in her fifties, a rather stern-looking woman who wore her hair severely pulled back and looked at the world through pince-nez. Many of the students regarded her with awe and were a little frightened and uncertain how they were supposed to respond to her in the laboratory. But Lilian was completely committed to the *Drosophila* work and good at it, and slowly the students recognized her basic kindness and generosity. Unlike her husband, who had the southerner's easy charm when he wanted to put it on and could fall back on his reputation for impishness and abruptness when he didn't, it took Mrs. Morgan a little while to warm up to people. But she virtually mothered some of the wives of the students, especially those from other countries, and proved a good friend to many.

Despite her tenuous connection with Fly Room, she was an active part of the give and take and continued disorder of that strange place. One day she realized that the fly under her microscope was a new mutant. Suddenly the fly was gone. Everyone began scouring the floor, thinking it had been knocked off the glass slide, but Lilian decided it must have recovered from the anesthesia and have flown away. And since *Drosophila* tend to seek the light, she went to the window and spotted it right away—a more difficult task than it sounds, since the room was always filled with dozens of escaped flies and even strangers drawn to the fermenting bananas and half-filled garbage cans.

The fly proved to be a most unusual female because its progeny violated the usual criss-cross pattern of sex-linked inheritance. Ordinarily, as with color-blindness in humans or white eye in *Drosophila*, the sons inherit sex-linked genes from their mother and the daughters inherit from their father or their mother. But when this wandering female fly which showed yellow abdomen, a sex-linked recessive characteristic like white eye, was mated with a normal male, all the *daughters* showed yellow abdomen like their mother, and all the *males* resembled their father. This was just the opposite of what had been expected. The best explanation was that the mother's chromosomes were XXY. The two X chromosomes were joined together, and so two types of ova were produced: some XX and some Y. These in turn might be fertilized by either an X- or a Y-bearing sperm. The YY individuals perished and the XXX—superfemales—were feeble and died readily. The remaining majority were females (XXY) that received no X from their father, and males (XY) who did receive an X from the father and hence his X-linked genes. Cytological investigation confirmed the breeding results.

This experiment also confirmed Bridges's balance theory of sex determination that it is not just the presence of X or Y chromosomes that determines sex but the final balance at fertilization between the Xs and autosomes. Mrs. Morgan wrote a friend excitedly:

The fly was a female with yellow abdomen. Apparently at some division the halves of the paternal yellow chromosome became inseparable, producing a mosaic and never became separated through (up to now) three generations. This is based on the fact that all the males are like their father whatever the mating may be and all the female offspring in three generations are yellow. In the F_1 generation the males were infertile, as would be expected if this explanation is correct. A few ♀ occur that are like Mr. Bridges 3XXX flies in the points where they are slightly "abnormal" and these show the color expected with yellow and an X from the father.

In the 1920s, discoveries of new gene mutants or new chromosomal mutants became less frequent. Much of the work in this line had come to an end. Certainly the questions that had brought Morgan to genetics had been answered to his satisfaction, and his usual prolific output of articles and books took on the tone of wrapping up and summarizing the genetic investigations of *Drosophila*.

During these last years at Columbia University, Morgan was receiving many of the rewards of his successes: new editions and translations of his books, honorary degrees, and high office, such as the presidency, in 1927, of the National Academy of Sciences. But even higher evidence of the respect he had won was the stream of scientists from all over the world who were drawn to Morgan's laboratory. Some came to confer, as did Landsteiner and Levine in 1927 when the second blood group MN was discovered and they wanted assistance with working out the inheritance. Others simply wished to "gaze on the wonders of Columbia," as William Bateson wrote in 1922 after his trip from England to the Fly Room. The rightness and validity of Morgan's experimental results were recognized in event after event, although perhaps no conversion was so dramatic as that of Bateson, who, until he actually came to the Fly Room, disbelieved that the gene had a physical reality as part of the chromosome. Clearly exceptions are easier to treasure if they are your own. Afterwards Bateson went to Toronto to address the Third International Genetics Congress, an account later published in *Science* 50 (January 20, 1922):55–61.

The arguments of Morgan and his colleagues, and especially the demonstrations of Bridges, must allay all scepticism as to the direct association of particular chromosomes with particular features of the zygote [fertilized egg].

The traces of order in variation and heredity which so lately seemed paradoxical curiosities have led step by step to this beautiful discovery. I come at this Christmas season to pay my respectful homage before the star that has arisen in the west.

In the same presentation, Bateson's qualified opposition to Darwinian evolution—and the press's unbalanced account of it as an eminent geneticist's disbelief in Darwinian evolution—refueled the evolution-fundamentalist debate in America. This made Morgan indirectly responsible for the problems many states would face. Tennessee had its famous Scopes monkey trial and Kentucky in 1932 avoided fundamentalist laws by a narrow margin. (A bill was introduced in the Kentucky legislature forbidding the use in the public schools of textbooks that included the doctrine of evolution. The first vote, 38 to 36 for the bill, failed because 40 votes were needed for passage. Absentees were called in by the proponents. A second vote was 40–39 for. Opponents demanded a halt and recount. Three more people dragged in, making the next vote 41–41. Representative Cunall, who had declined to vote previously because he was "a hard-shell Baptist and believed what was would be anyhow" then said he'd have to discard his religion and vote no. The bill was defeated. Morgan seems to have noticed neither the furor nor the finish.)

Whoever came to Schermerhorn Hall found Morgan the same. Levine admitted some astonishment at finding the great man eating his lunch in the Fly Room, "with his elbows close to side as there was so much clutter around." A visit to the Fly Room was high priority for graduate and postdoctoral students in genetics from all over the world. It was often not quite what they had expected. Theodosius Dobzhansky, born and trained in Russia, arrived breathless in New York, expecting the eminent Professor Morgan to appear "like a god." The newcomer was taken back by the abrupt and always carelessly dressed man who observed none of the formalities and would not even wear a lab coat (although he did once, for the *Vogue* photographer assigned to do a series on Nobel laureates at Caltech). Dobzhansky stayed with Morgan from 1927 to 1940, when he returned from the West Coast to Columbia as Morgan's successor. Morgan never could or would pronounce his name correctly.

It was also difficult initially for some of the scientists to adjust to the autonomy expected of workers in the Fly Room. Morgan built up a group of workers in whom he believed, and he expected them to know what they wanted to do and how they wanted to go about it. He then left them alone to do it—except that he could never entirely suppress his enthusiasm for any experiment and would dart in, peering over shoulders, making on-the-spot suggestions, or asking questions. He always was a coordinator of all the work going on, but his most important role in the Fly Room the last few years may well have been what his "student" A. H. Sturtevant called his "unfailing support, stimulus and protection" to the workers in his domain.

By 1927 the *Drosophila* work had been going on for some eighteen years—about 15,000 generations of flies from Dr. Payne's first strain. Morgan had given over much of it to Bridges and Sturtevant (whom he continued out of habit to call his boys though their Ph.D.'s were twenty years behind them and their contributions to science firmly established). He himself was in spirit and in approach boyish, and as Sturtevant later wrote, "at his best in opening up new and little understood fields. He appreciated the value of detailed and analytical studies in better known fields, and was always ready to support such work, but he himself was usually off on some new . . . line. He always had many types of work under way—a fact that is only partly shown by the diversity of subjects on which he published, for many of his projects did not yield significant results, and were never recorded." One such unrecorded experiment was, in 1924, still Lamarckian. Mice were bred for two years to see if subsequent generations would inherit a conditioned reflex, as Pavlov claimed. The outcome was negative.

Morgan maintained his lifelong interest in natural history. He was able to identify almost any specimen given him. In 1920 Konrad Lorenz, then a teenager but later to become the founder of ethology, met the geneticist after being told that a Professor Morgan up in Schermerhorn Hall could help him identify a specimen that no one else at Columbia could. Morgan—whom Lorenz recalls looking as tall and spare as Abraham Lincoln—was pleased to identify this and any other creature Lorenz brought him and patiently explained the biology, morphology, and ecology of each. It was typical of the generosity Morgan always showed to those he felt deserved it that he also carefully explained his work with the tiers upon tiers of little bottles filled with flies and showed this unknown teenager his first chromosomes under the microscope.

In 1927 Morgan had been at Columbia University for almost a quarter of a century and, at the age of sixty-one, was nearing compulsory retirement from teaching. He clearly had accomplished half his life's ambition, but, as always, his experimental work consumed the bulk of his interest. There were important questions still unanswered; uppermost in his mind was the problem of development. It was not yet known how genes caused development, or even if genes were responsible. Many a man would have adjusted his laurels and viewed retirement as time to continue his work in the peace of a lovely home. Not Morgan. He took up the correspondence that was to lead him at sixty-two to California—heaven's half-way house—there to establish an ideal school of biology. He went, incredibly, with the stated intention of reaching for the other half of the apple of life, the answer to the question of differentiation—how an undifferentiated cell, the ovum, produces daughter cells that become specialized parts of an adult.

8

CALTECH

Tom is a great man of science and knows everything about everything except why a chicken's egg don't turn into a crocodile.

Charles Kingsley
The Water Babies

IN 1928, at a reception given for him as the new head of the biology division at the California Institute of Technology in Pasadena, Morgan explained, "Of course I expected to go to California when I died, but the call to come to the Institute arrived a few years earlier, and I took advantage of the opportunity to see what my future life would be like."

The call was to organize and direct a completely new school of biology. California Institute of Technology had grown out of the nineteenth-century Throop Polytechnic Institute, a manual training school. During the years after World War I, especially after physicist Robert A. Millikan was persuaded to become its administrative head, the newly named California Institute of Technology developed rapidly and solidly from a good engineering college into a world center of research in the physical sciences. Millikan himself received the Nobel Prize in physics in 1923. He was also a very handsome and charming man who accumulated numerous large donations for the school. By 1927, physics, chemistry, geology, and aeronautics had joined engineering as fields in which top men were concentrated, the best graduate students enrolled, and important research conducted. Morgan was invited to organize the next division, that of biology.

In July 1927, Morgan took the job, sight unseen. He asked to be allowed another year at Columbia, however. He did not want to leave Wilson hanging with an empty department to be organized—especially as he planned to scoop off the cream for Caltech. And during this last year in New York, he set about making plans for the new school.

Accepting such a large-scale administrative position had its ironies for Morgan. He had always referred to himself as "a laboratory animal who has tried hard most of his life to keep away from such entanglements," and he seemed a little worried that some of his friends would think he had made a foolish mistake.

It was true that he himself obviously preferred working alone or in the clubbish atmosphere of the Fly Room. His antipathy to large-scale joint enterprises was well known. When his daughter Isabel, later a microbiologist with the National Foundation for Infantile Paralysis team, would argue that it was hard for an individual to do valuable research any more, Morgan always scoffed and accused her of "that Rockefeller Institute mentality." But the position at Caltech also meant the chance to build a school of biology to his own specifications. Research, not teaching, would be the major focus, and the research would be primarily "pure," rather than tied to any immediate practical application. The faculty and graduate students would be Morgan's ideal: people who knew what they wanted to do and who set about doing it with a minimum of guidance and a minimum of speculation. Furthermore, building this program within a well-established physical sciences research institution meant that the work would be rigorous and analytical, so that biology would be conducted by the same standards as work in physics and chemistry.

Morgan made plans along these lines, justifying such

omissions as morphology by the rationale that Caltech didn't want to duplicate what was available elsewhere— at Hopkins and Columbia, for example—but to pioneer new directions of research and education. What he was doing was setting up the kind of place *he* had always wanted, a grand mixture of the best of the Fly Room and the Naples station. But a more remarkable emphasis, in that he did not particularly care for it in his own work, was on the cooperation of biology with sciences such as physics and chemistry. The biology complex at Caltech was even planned to touch the chemistry wing in order to provide physical contact and continually encourage collaboration with other basic sciences. Botany, zoology, and genetics were brought together under one department of biology (though Morgan rejected the idea of building a hospital). As he wrote in his initial plans, "An effort will be made to bring together, in a single group, men whose common interests are in the discovery of the unity of the phenomena of living organisms." His own understanding of the concepts of mathematics, physics and chemistry was limited, as he himself realized, but his plan was farsighted—and his collaborative program well realized in the work of the men he set about recruiting for the new faculty. The nineteenth-century idea that biology could be explained with physics and chemistry was beautifully expressed by D'Arcy Thompson: "Cell and tissue, shell and bone, leaf and flower, are so many portions of matter, and it is in obedience to the laws of physics that their particles have been moved, molded and conformed." Morgan always endorsed this view: "We realize that only through an exact knowledge of the chemical and physical changes taking place in development can we hope to raise the study of development to the level of an exact science." It was clear that he did not have any specific approach in mind so much as admiration for the experimental method that had earned physics and chemistry such success.

Morgan himself could not have been expected to

abandon his faith in learning by experiment and enjoy theoretical speculation with a Schrödinger or a Pauling.[1] Their two different approaches were well expressed by Linus Pauling, who wrote: "At that time (1937) I found that Landsteiner and I had a much different approach to science: Landsteiner would ask, 'What do these experimental observations force us to believe about the nature of the world' and I would ask, 'What is the most simple, general and intellectually satisfying picture of the world that encompasses these observations and is not incompatible with them.' " Morgan was entirely on the side of Landsteiner; he believed that truth came only from experiment. But in the division of biology Morgan was building at Caltech, the two approaches would meet.

Morgan had fairly firm ideas about the areas that should go into the new biology division: genetics and evolution; experimental embryology; general physiology; biophysics; biochemistry. Others, such as psychology, could be added later. But the exact shape and direction of the division and its departments would depend on the faculty. Morgan did not intend merely to fill empty slots, but to accumulate, as slowly as necessary, the best people available and let them follow their own interests and strengths. For this reason, although he was seeking advice about several of the proposed areas, the only department to operate during the first year (1928–1929) was genetics. From the department of zoology at Columbia he recruited Bridges, Sturtevant, Jack Schultz, and Albert Tyler. From there he also took on Theodosius Dobzhansky, first as an International Research Fellow, the following year as an assistant professor. By 1931 there were the following staff members besides the Columbia team: Ernest G. Anderson, Henry Borsook, Herman Dolk, Robert Emerson, Sterling Emerson, Hugh Huffman, Kaj. Linderstrom-Lang, Henry Sims, and Kenneth V. Thimann. James Bonner and Herman Schott were graduate students and George Wells Beadle was a postdoctoral student; Geoffrey Keighley was an assistant to

Henry Borsook; Walter Lammerts was a postdoctoral fellow with Professor Anderson.

During his last year at Columbia, Morgan had worked with the blueprints for the new buildings at Caltech, which were partly constructed when the small faculty, half a dozen graduate students, and a large stock of *Drosophila* arrived, via Woods Hole, in the late summer of 1928. They used what space they could and camped out for a few months in other people's offices and classrooms, while Morgan oversaw completion of the building and organization of the library and laboratory equipment.

Having to buy that much equipment at once was difficult for a man so notoriously stingy with institutional money that he destroyed all his own files every five years to avoid buying another file cabinet. The modern biology complex at Caltech had one telephone on each floor and one secretary for the whole building. A scientist who came asking Morgan for another case of the shell vials for *Drosophila*, valued at something under a cent each, spent the next two hours going through the laboratory with the Father of Genetics, who was certain they could find enough old ones. George Beadle recalls carefully laying plans to find Morgan at his embryology laboratory at the shore, on a Sunday when he was alone with his experiments and thus happiest, so that he would be assured of a good moment to ask for a new $90 microscope objective, which to general surprise he got.

While it is not possible even to summarize the vast amount of important work that was begun at Caltech, some of George Beadle's work will serve as an example. When he joined Morgan at Caltech as a postgraduate fellow, he began to work on crossing over in *Drosophila*. In 1933 he and Ephrussi, who was visiting Caltech that year, realized that Sturtevant's report in 1919 of a vermilion-eye mutant that reverted to normal eye color in a gynandromorph was a possible clue to the method of gene action. Since the effect of the genes in the eye appeared to be modified by the surrounding tissues, they wondered if the problem of gene action could be resolved by studying a variety of eye mutants in a variety of environments. They attempted to surgically remove the eye bud of a mutant larva and transplant it into the body cavity of a wild-type larva. When the host larva underwent metamorphosis and became an adult fly, the implanted parasitic eye could be removed and its color observed. Beadle applied to the Rockefeller Foundation for support to enable him to go to Paris with Ephrussi for one year to learn the technique of organ transplantation, but he was turned down. At this point the requisite $1,800 suddenly came from Caltech, either through Morgan's administrative direction or, as Beadle believes, from Morgan's pocket. Either way, it was a good measure of Morgan's nose for important work and his determination that it should proceed. Shortly after Beadle arrived at the Institute de Biologie Physiochemique, a noted authority on the metamorphosis of the blowfly told the two that their proposed work on *Drosophila* transplantation was quite impossible.

They succeeded, however, and found that while most mutant eye buds developed into their predetermined mutant colors, both the vermilion and cinnabar buds changed to the normal red eye color "not by the genetic constitution of the eye pigment itself but by that of some other part of the body." They reasoned that some substance had diffused from the wild-type tissue to the eye, causing it to become naturally pigmented. On the basis of reciprocal transplantations, the only possible explanation was the existence of a sequence of gene-controlled metabolic steps, each one of which was mediated by a specific enzyme.

	↓ Enzyme 1	↓ Enzyme 2	↓ Enzyme 3
Precursor →	vermilion substance	→ cinnabar substance	→ normal pigment

117

On his return from Paris, Beadle went to Stanford, where he and Tatum tried to identify the eye-color substances chemically, but failed. The two then hit on the remarkable idea that instead of investigating whatever mutants happened to be available, it might be better to choose an ideal organism and induce the mutations that were required. Accordingly they decided to use the fungus *Neurospora crassa* that Morgan had brought with him to Caltech from Columbia, where he had been persuaded to adopt it by the astute and persistent Dr. B. O. Dodge of the New York Botanical Garden. Dodge had been struck by failure of *Neurospora* to arrange its ascospores in accordance with prevailing genetic theory and urged Morgan to investigate it as a preferable organism to *Drosophila.* Morgan eventually agreed and gave some contaminated cultures to a young graduate student, Carl C. Lindegren, who proceeded to work out much of its genetics. Beadle and Tatum saw it as an ideal organism because: (1) its genetics were well understood, thanks to Lindegren; (2) irradiating spores with X-rays or ultraviolet light readily induced mutations; and (3) the fungus would grow on a defined minimal medium and synthesize all its requirements; thus biochemical mutants were easy to identify by their failure to grow. The defect could then be identified by finding which substance had to be added to stimulate growth. Irradiation induced 380 mutants, which were then mated, and over 68,000 single ascospores were examined. Eventually this work led to the concept that genes are responsible for the synthesis of enzymes which control specific biochemical reactions. In 1958 the study resulted in the third Nobel award to genetics. (The second had gone to Muller in 1946.)

California proved itself to be the paradise Morgan envisioned, and in a short time he had eased himself into a pleasant pattern of life. It was only Tom and Lilian again; the children had all grown up. Howard had finished at the University of California after a false start, married a fellow student named Bernadine Buck, and was in the process of becoming a successful engineer. Edith graduated from Bryn Mawr the summer of 1928 and married Douglas Whitaker, a scientist (later to become a scientific administrator) and family friend. The younger daughters Lilian and Isabel finished high school that same year and had chosen California colleges—Isabel going to Stanford, Lilian to Pomona. They were in and out of Pasadena whenever they could be—almost every weekend for Lilian since Pomona was nearby and somewhat less frequently for Isabel since Stanford was further up the coast—and the family, including the first of the grandchildren, met most summers at the Morgans' large house in Woods Hole.

Tom and Lilian bought a beautiful old ranch house, built by a family of Spanish landowners, at 1149 San Pasqual Street. All the elegant mahogany furniture shipped from the house in New York seemed to take on a more relaxed look in the large, sunny rooms of the California house. The player piano from the New York house had bumped its way around the southern coast, somewhere fatally disrupting its mechanism, but the Morgans set it up in the new house anyway, like an old friend. The junior-size billiard table of New York was replaced with a full-size one. The biological laboratory was only across the street, and Caltech bought part of the large lot upon which the house itself sat and built a greenhouse, so that Morgan continued his habit begun in New York of stopping to pick a perfect red rose for Lilian at the end of the day. In the mornings they both set off to the laboratory to work, although Morgan now had to divide his time between research and administration with the latter winning most often. At lunch they met at home. Great beams stretched across the living room, and the house curved around a patio where they sat in the sun after eating, Morgan reading and puffing on a fifteen-cent Bobby Burns cigar.

And there he sat, calmly reading *Anthony Adverse* (the

year's most popular adventure story), as if nothing much had happened, Lilian wrote one of her daughters in 1933. What had happened was the telegram announcing that, on the hundredth anniversary of Alfred Nobel's birth, Thomas Hunt Morgan had been awarded the Nobel Prize for his work on the chromosome theory of inheritance.

The Nobel award came as a surprise. Morgan had been nominated twice before for the same work: in 1919 by Ross Harrison and in 1930 by Dr. Otto Mohr who was by then Rector of Oslo University. Mohr used to say that these nominations were turned down on the grounds that genetics was not physiology or medicine. With two exceptions, recipients of the prize in this category had always been medical men or members of medical faculties. Morgan, of course, was neither, although in 1933 the University of Zurich did give him an honorary M.D. The third nomination of Morgan was by Karl Landsteiner, the immunologist, physician, and Nobel laureate who had earlier sought Morgan's help (although he had gotten mostly Bridges's and Sturtevant's) in working out the inheritance of the MN blood group that he and Levine discovered in 1927.

Despite his obvious though somewhat embarrassed pleasure in such public signs of success as being in the newsreels at the movies, Morgan was very modest about the honor. He frequently pointed out that it was a tribute to experimental biology rather than any one man. It seems that the prize might have gone to the entire Fly Room team, if it had not had four members. (The award may not be shared among more than three.) As Morgan acknowledged the joint nature of his work, he divided the tax-free $40,000 award equally among his own children and those of Bridges and Sturtevant (but not Muller's). He gave no reason; in the letter to Sturtevant for example, he said merely I'm enclosing some money for your children. (Bridges, however, is said to have used his to build a new car.)

Morgan passed up the opportunity to attend the lavish

Nobel banquets held on December 10, Nobel's birthday, in Stockholm. His excuse was that "circumstances here in connection with the establishment of a new group in physiology and with the immediate future of biochemistry in genetics make it imperative for me to remain." Doubtless another factor was his longtime dislike of fancy dress occasions. A probable third was the rediscovery of the giant chromosomes in the salivary glands of the larvae of *Drosophila* and other flies. This phenomenon that nature had obligingly provided for genetics—the 2,000-times magnification of the chromosomes—had been reported by Balbiani in 1881 and then forgotten. The rediscovery and announcement of these chromosomes, published by Heitz and Bauer in January 1933 and by Painter in December of the same year, came at a very inconvenient moment for writing, much less delivering, a Nobel acceptance speech. Most of the Morgan school's contributions to the chromosome theory of inheritance were actually inferences, based on genetic studies with little direct help from the chromosomes; even the cytological evidence for crossing over was not certain. But now, instead of struggling to interpret minute or invisible changes in the small and undifferentiated metaphase chromosomes of *Drosophila*, scientists could read them from the large and clearly visible segments of the salivary gland chromosomes. These large chromosomes showed numerous fine transverse bands that provided an independent way to verify or refute the linkage maps and postulated deficiencies, duplications, and inversions. It was the test by fire of the Morgan school's work. Would the new data refute or confirm the chromosomal results that had been predicted? And could Morgan maintain his preeminence?

Morgan told the Nobel Committee that he would be pleased to come next summer. Morgan and his wife left for New York in April of 1934 and then, with daughter Isabel, sailed to London, visited the Mohrs in Oslo, and went on to Stockholm for the special ceremonies held for

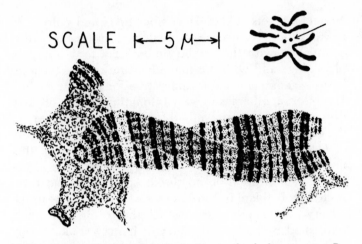

SCALE ⊢—5 μ—⊣

Figure 8. Photograph of a giant salivary gland chromosome. Reprinted, by permission, from Bridges, "Salivary Chromosome Maps," *Journal of Heredity* 26 (1935), figure 4

Morgan. By this time the results were in. Morgan's work was confirmed, although there were details still to be settled. Accordingly, he devoted much of his speech to the discoveries of the preceding twelve months, although he did not include in the original speech—delivered in June 1934—the linkage maps. He may still have feared that they would not prove accurate.

The speech itself was not published in America until July 1935, when it appeared in a relatively obscure (now defunct) journal called the *Scientific Monthly* (41:5–18). In comparing the American version with the original speech, it is interesting to see what Morgan inserted: he added the linkage maps, about whose accuracy there was now no doubt (though in neither edition did he mention Sturtevant), and new plates of giant salivary gland chromosomes, including the marvelous fourth chromosome, published in February 1935 (given here as Figure 8). Whether Morgan intended it or not, the inclusion of this material made it appear that Heitz, Painter, and Bridges had been running neck and neck in their work on giant salivary chromosomes back in 1933, when Painter was clearly ahead. In fact, Dr. Robert Olby, the careful author of an excellent history of genetics, assumed that scientists like Caspersson got their first glimpse of visible genes from the Morgan plates presented in Stockholm in 1933. They did not.

The speech, entitled "The Contribution of Genetics to Medicine and Physiology," deals with medicine half-heartedly. Morgan clearly saw little contribution aside from genetic counseling. Although he wrote the speech near Dr. Albers Fölling's laboratories at a time when Fölling had just discovered PKU and was working on its biochemistry, Morgan made no mention of it nor of biochemical genetics. And he made no mention of Adrian Bleyer's marvelous inference (made in 1932 and ignored till 1959) that Down's syndrome was due to nondisjunction.

Morgan shared his disinterest in medicine with most American geneticists, who were initially opposed to the postwar development of a society and journal of human genetics. This disinterest in medicine was not just theoretical for Morgan, for when his daughter Isabel had discharging tuberculous neck glands, Dr. Tove Mohr (who had only been qualified as a physician one month) failed to persuade Morgan that the child should see a doctor—"Oh it's all right, you are our doctor," he argued.

The speech, however, does contain fascinating speculations about the possible mechanisms of gene control.

If, as is generally implied in genetic work (although not often explicitly stated), all the genes are active all the time; and if the characters of the individual are determined by the genes, then why are not all the cells of the body exactly alike?

The same paradox appears when we turn to the development of the egg into an embryo. The egg appears to be an unspecialized cell, destined to undergo a prescribed and known series of changes leading to the differentiation of organs and tissues. At

123

every division of the egg, the chromosomes split lengthwise into exactly equivalent halves. Every cell comes to contain the same kinds of genes. Why, then, is it that some cells become muscle cells, some nerve cells and others remain reproductive cells?

The answer to these questions seemed relatively simple at the end of the last century. The protoplasm of the egg is visibly different at different levels. The fate of the cells in each region is determined, it was said, by the differences in different protoplasmic regions of the egg.

Such a view is consistent with the idea that the genes are all acting; the initial stages of development being the outcome of a reaction between the identical output of the genes and the different regions of the egg. This seemed to give a satisfactory *picture* of development, even if it did not give us a *scientific explanation* of the kind of reactions taking place.

But there is an alternative view that cannot be ignored. It is conceivable that different batteries of genes come into action one after the other, as the embryo passes through its stages of development. This sequence might be assumed to be an automatic property of the chain of genes. Such an assumption would, without proof, beg the whole question of embryonic development, and could not be regarded as a satisfactory solution. But it might be that in different regions of the egg there is a reaction between the kind of protoplasm present in those regions and specific genes in the nuclei; certain genes being more affected in one region of the egg, other genes in other regions. Such a view might give also a purely formal hypothesis to account for the differentiation of the cells of the embryo. The initial steps would be given in the regional constitution of the egg.

The first responsive output of the genes would then be supposed to affect the protoplasm of the cells in which they lie. The changed protoplasm would now act reciprocally on the genes, bringing into activity additional or other batteries of genes. If true this would give a pleasing picture of the developmental process.

One of Morgan's reasons for leaving Columbia had been impending and mandatory retirement; a reason that the head of Caltech understood. He himself was over sixty when he hired Morgan, and he planned to and did spend another quarter-century in research and administration. The original agreement was that Morgan would serve as director of the new division for five years, ending in 1933, at which time he could retire or the board of trustees could retire him. But when the time came, Morgan wanted to continue the work at Caltech. He was not satisfied with the progress made toward development of the biology division. The board of trustees, however, was delighted with his work and, of course, with the increased reputation of the school, and it gave him the additional five years for which he asked. In 1938, when he was seventy-two, he asked for yet four more years, which he was given. Only in 1942, at the age of seventy-six, did Morgan retire to become professor and chairman emeritus.

Morgan was not replaced immediately at Caltech. Sturtevant ran the biology division as chairman of a biological council from 1942 until 1946. In 1946 George Beadle returned from Stanford to serve as chairman of the division until 1961, when he became chancellor of the University of Chicago.

After his retirement Morgan maintained his office and laboratory space in the campus across the street. In the place of Woods Hole, he had the Corona del Mar marine biological station that Caltech had bought and outfitted under his direction. On Sundays he was driven there, about an hour's trip, by one of the younger teachers, but he worked in his Caltech laboratory the other six days of the week.

The *Drosophila* work—the world center of which had moved to Pasadena along with Morgan—was carried on largely without the man who began it. When the time came for a new summary of genetics, it was produced by Sturtevant and Beadle and called *An Introduction to Genetics*. It is still an excellent book and still in print. In a review written shortly before his death, Haldane called it the best introduction to classical genetics available. Sturtevant, Bridges, and many others in the Caltech genetics department, even Morgan's wife, pursued *Drosophila* experiments as intensely as ever, while Morgan gave his energies largely to other types of work. Perhaps his willingness to do so could, like his decision to become an administrator, be partly explained by his feeling that the *Drosophila* group had made its most important contributions to genetics and that the new work to be done lay in directions such as physics and chemistry and population genetics.

Although he set the stage for exactly this kind of work in the Caltech program, Morgan did not always appreciate it. Curt Stern has told of the time the geneticist Bernstein gave a seminar at Caltech in which he demonstrated, mathematically, that there were three alleles at the ABO blood group locus. After the talk, Morgan asked whether the same results could not have been obtained without the mathematics, by using pedigrees. They could have, Bernstein replied—but they weren't!

Nonetheless, with his almost unerring instinct, Morgan recognized the value of what he did not always understand or even like. Caltech continued to attract the ablest scientists throughout the world. For example, Max Delbrück joined Caltech in 1937 and later wrote: "I chose Caltech as the place to go to because of its strengths in *Drosophila* genetics, its sympathetic attitude towards my scientific interests in general." While Bridges and Sturtevant continued to map *Drosophila*, and Beadle was beginning *Neurospora* studies and Delbrück, thanks to Emory Ellis, chose an even smaller organism, a bacterial virus, Morgan returned to many of the simple but much larger organisms with which he had begun his research at the turn of the century—and to the basic questions of regeneration and differentiation, in particular sexual differentiation. He investigated during this period the seasonally developed secondary sexual characteristics of the salamander *Triturus*, regeneration of the arms of brittle stars, crosses between different geographical types of deer mice, and rapid changes in color of the living "goldbug," *Coptocycla.* His last experiments, resembling in approach his first, involved what he referred to as his "blessed *Ciona.*" One day, discussing the problem of self-sterility in *Ciona,* Morgan said that acidified sea water might overcome the block to self-fertilization. It was suggested that weak acids might be better for this purpose than strong ones. As no weak acids were available in the laboratory, Morgan took a lemon from his lunch box and squeezed the juice in measured portions into dishes of sea water containing *Ciona* eggs. The work he was doing when he died was a continuation of his old investigation of self-sterility in *Ciona.* This resulted in the discovery of the existence of pairs of genes for self-sterility in more than one pair of chromosomes, achieving their effect by producing a protein coat on the ovum.

Morgan continued to publish numerous papers, but he published his last two books in the early 1930s. *Embryology and Genetics* tended to treat the two subjects somewhat separately, as was inevitable at that time since *Ciona* was not suitable for genetic studies and *Drosophila* did not lend itself to embryological investigation. In 1932 he published *The Scientific Basis of Evolution,* which contained not a single line of mathematics, despite the recent work of Haldane, Fisher, and Wright, who applied rigorous quantitative methods rather than descriptive morphology to the problem.

Morgan devoted a chapter to the inheritance of acquired characteristics but explained: "It is somewhat depressing to give so much time to destructive criticism of a doctrine that makes wide popular appeal. It sometimes seems as if everybody wanted to believe in the inheritance of acquired characters. There is a mystery about it which gives it an emotional setting. Yet if we are not to remain the dupes of our emotions, it is part of the role of science to destroy pernicious superstitions, regardless of the appeal that they make to individuals inexperienced in the exacting methods that science demands." He noted that "there are, in man, two processes of inheritance; one through the physical continuity of the germ cells; and the other through the transmission of the experiences of one generation to the next by means of example or by spoken and written language." He then shied away from eugenics, extolling cultural inheritance: "The doctrine that all men are born free and equal carries with it the assumption that they should be allowed to breed as they will."

In the second edition, which appeared in 1934, he added an extra chapter in which he included half of his Nobel Prize speech even though it had nothing to do with evolution.

Morgan had always been healthy. In the fall of 1931, when he was 65, an automobile collision caused a piece of windshield glass to penetrate his back. A passing medical student named Leon Baker held the glass in place in the wound to slow the hemorrhaging. (When Baker found himself in financial difficulty later, he was given an anonymous fellowship, mysteriously resembling the anonymous awards sometimes given Columbia biology students.) For two months Morgan experienced pain, bleeding, and the miseries of enforced rest. He was able to go back to work only after the first of the year, and even then he had to restrict his activities for several more months. No one suffered more from such interruptions than Morgan, even though they were few.

Morgan had only one chronic illness and that was a duodenal ulcer. All his troubles went to his stomach. The few deadlines he had to meet bothered him. So did giving lectures. He was always being invited to lecture and generally refused, despite high fees. Once when he was to lecture at Stanford he was quite nervous. His son-in-law Douglas Whitaker offered him a drink but could only find an eight-ounce glass. Whitaker began to pour, instructing Morgan to say "when." Morgan never did, and so, to Whitaker's consternation, tossed back almost eight ounces of whiskey. He assured the younger man there was no need to worry about him ("I've got myself calibrated") and went on to give a fine lecture.

While Morgan never complained about any tensions, nor about any aches and pains, Lilian judged the day at the university by his appetite and measured his different books by what they cost in nervous stomach upsets. Like so much in the Morgan family, his tensions and their intestinal expression were never discussed but were understood.

In 1945 Morgan experienced his most severe attack. He took it jokingly and refused to admit either pain or the possibility of serious medical problems. But when he began hemorrhaging in November he was admitted to Huntington Memorial Hospital in Pasadena and the children were called home. On December 4, he died from a ruptured artery. His body was cremated. There was no public ceremony, only the gathering of a few friends in Pasadena and a similar gathering held in New York.

Lilian was stoical about his death. She spent some time visiting each of her children. Both Howard and Edith had families, and Edith had returned to school to become a physical therapist. The second daughter, Lilian, had since married Henry W. Scherp, another scientist, and was herself a medical social worker. Isabel, the youngest, had completed her doctorate in microbiology and was then in Baltimore working with the polio project. During 1946 she demonstrated that it was possible to immunize a monkey against polio, a milestone in that it opened the question as to whether it was feasible to immunize humans. Her mother spent several months with Isabel but would not stay longer; she said she didn't want to interfere with Isabel's chances for marriage. (This daughter *was* later married, to scientist Joseph Mountain, and, as Mr. Mountain always said, to his young son Jim, whom Isabel subsequently adopted.)

Lilian returned to the house across the street from Caltech and continued her own laboratory work. In 1952 she became very ill, but ignored the fact until the children forced her to change doctors. Her illness was diagnosed as intestinal cancer, and it progressed quickly. Either she was not very susceptible to pain, or, like Morgan, refused to acknowledge it. She wrote her last article from a hospital bed. She did not believe in supporting a body that could not support itself, and let the illness go its way. When it became obvious she was only minutes from death, the doctor asked if there was anything she wanted to say. "No," she replied. "I think everyone understands."

9

CONCLUSION

Alice: I can't believe impossible things.
Queen: I dare say you haven't had much practice.
Lewis Carroll

Biographies inevitably concentrate on one individual's genius, while the development of ideas is vastly more complex and interwoven. The following list of achievements really belongs to the team of which Morgan was the undisputed Boss. Through a series of fascinating, audacious, yet simple experiments, the team:

1) demonstrated the physical reality of the gene as part of the chromosome;

2) confirmed Mendel's laws;

3) rediscovered 'linkage,' an exception to Mendel's laws, and crossing over, and double crossing over, exceptions to linkage;

4) discovered the linear order of genes whose relative position could be fixed and the distance between them measured (in morgans and centimorgans);

5) demonstrated that the determination of sex was chromosomal;

6) discovered duplications, deficiencies, translocations, inversions, trisomics, triploidies, and attached X chromosomes;

7) discovered the position effect, the multiple effects of one gene, multiple alleles, and single characters influenced by multiple genes.

It is not surprising that Morgan is remembered as a geneticist. He, however, thought of himself as an experimental zoologist whose main interest was experimental embryology. While Morgan was too busy with genetics to lay down the foundations of embryology as well, he did develop the theory of gradients in 1904 and 1905. He distinguished two processes of regeneration and coined the term morphallaxis. He discovered that the presence of nerve tissue was necessary for regeneration and showed that regeneration was not an adaptive phenomenon. He also investigated the factors influencing the ovum. He measured the correlation between the site of entry of the sperm, the planes of first cleavage, and the eventual planes of symmetry. He discovered that magnesium salts would initiate parthenogenesis and that lithium salts, low temperature, and anoxia, would initiate teratogenesis. He disproved the mosaic theory by the demonstration of the *normal* development of an isolated blastomere, and he understood long before anyone else that genetics would explain embryology only when some mechanism was proved to regulate gene activity.

Morgan even demonstrated that embryology was his highest priority by forsaking his normal rushed method of book production and instead spent over seven years of painstaking work writing *Experimental Embryology,* which was promptly ignored. Indeed most of Morgan's embryology is undiscovered. At the 1966 centennial symposium held in Morgan's honor, distinguished scientists read papers on genetics and development, but none of them quoted his embryological work, although their respect and admiration for him was obvious. The reason for this neglect is that Morgan was then (as the title of our book indicates he is now) clearly labeled a geneticist. The power of labeling is such that his opinions as a geneticist were accepted even when they were wrong, and yet his opinions as an embryologist were rejected even when they were right. Perhaps a similar phenomenon explains why Mendel's work was ignored (what does a monk know about science?) and why Garrod's brilliant discoveries lay fallow for so long (or a physician about genetics?).

Morgan was against labeling, and this attitude played a large part in the startling success of Caltech. There he discouraged the development of discrete departments and encouraged both the fusion of genetics, zoology, embryology, and physiology into biology and the blending of biology with chemistry and physics. This is all the more remarkable considering that Morgan liked to work alone with simple equipment and had little background in mathematics or physics or chemistry.

In his own assessment of the ingredients for success Morgan mentioned luck, a favorable organism, skepticism, and industry. He might have added the ability to keep important questions in mind and reject trivia. And he might have added his own willingness to tackle the impossible. When Driesch found that a machine (the ovum) could divide in half, each half having the same potential as the whole, he found the results impossible to believe, invoked a *deus ex machina*, and gave up embryology. Morgan also found the results impossible—and took up embryology. Another example of Morgan's attitude was perfectly captured in an exchange at Woods Hole with the young embryologist James Neel. In discussing the transplantation of the corpora allata, the ring gland of *Drosophila*, Neel said it would be technically very difficult; and Morgan answered with one of his own mottoes: "Difficult yes, but not impossible."

But there is another ingredient, one that we hope we have conveyed in this small book: the charm and force of his personality. In every conversation about Morgan we have had with people who knew him, we were told, "I liked Morgan." They liked his lack of pomp, the outward signs of which were his uncombed hair and his pants held up by string; they liked his willingness to tackle the impossible and his refusal to be overawed by the opinions of others; they liked his democratic approach to students and colleagues, whom he treated without regard to their credentials or personal lives; they liked his sense of humor, his boyish enthusiasm, and his joy in the give and take of the laboratory. They even seemed to like his incredible stinginess with institutional and public money, which contrasted with his personal generosity with his own money and time. Such is the stuff of legends. He is lastingly endeared to generations of scientists who passed through his laboratories: so it is no wonder that the house George Beadle bought from Morgan in Pasadena was always called the Morgan house, and when in 1948 a colleague referred to Beadle as the Boss, he was rebuked. Only Morgan was the Boss.

Some Europeans found Morgan intellectually undistinguished, but that, we think, is because they were not used to an American style with a Kentucky accent. The English geneticist William Bateson said, as Morgan noted in his obituary of Bateson: "Democracy regards class distinction as evil; we perceive it to be essential." It doubtless surprised Bateson to find a laboratory where students could argue as equals, and it appalled him to find it so smelly. Shortly after Bateson returned in 1922 from his Christmas trip to the Fly Room, Professor E. B. Ford met Charles Darwin's son Leonard in Picadilly, and they decided to look up Bateson and hear his news. They found him in the posh Atheneum Club. He told them that Morgan was right and all his own life's work had been in vain. Later, however, he returned to his earlier skepticism because Morgan's laboratory was "absolutely filthy."

When Julian Huxley wrote to the American Philosophical Society about his reminiscences of Morgan he said: "After his death on one of my lecture tours in the U.S.A. I took the occasion to visit his birthplace and ancestral home [an expedition organized by Dr. Leland Brown] and found it most interesting. It explained a great deal in his character." Huxley also mentioned that he cherished a copy of *The Physical Basis of Heredity* inscribed, "To Julian (The Apostate) from Thomas (The Heretic)."

We share Huxley's opinion that Morgan's character is easier to understand in Kentucky. Morgan's family remained ever stunned, though enthralled, by the Civil War. In particular his uncle, the star of Lexington, the hero of Kentucky, the Thunderbolt of the Confederacy, remains a figure in the background all of Morgan's life. It was a figure from which he never escaped, even in death, when a Lexington obituary referred to the Fly Room team as being not unlike John Hunt Morgan's men, those Confederate raiders of whom Thomas Hunt Morgan's father was one, of whom Morgan himself never spoke, neither to his family nor to his friends. Having an uncle of such heroic scale must have contributed to Morgan's tendency to take on all comers without hesitation.

Public affairs, however, ruined Morgan's immediate family. He must have felt that his father had paid the price of war, and he was close enough to it himself to know that Sherman spoke the truth when he said that "war is hell and its glory all moonshine." At the wild celebration of the 1918 Armistice while the Fly Room team went downtown to burn the Kaiser in effigy, Morgan stayed in the lab. When his friend Ross Harrison asked him to sign a petition to get Mrs. Harrison and other German civilians, including the geneticist Richard Goldschmidt, released from prison, Morgan refused. When Kentucky came within one vote of barring the teaching of evolution in the schools, Morgan did not add his name to the list of distinguished men wiring their support for the beleaguered president of the University of Kentucky.

Morgan was a conservative who did not see these things as his concern and made no pretense to. To what he did see as his concerns—the advancement of experimental biology, the pursuit of pure research, the steering of young scientists in the right direction—he gave his all. His dedication to enquiry made him a beacon of scientific truthfulness at a time when mystics, fanatics, and fools were not uncommon. His belief in only those hypotheses that were based on observation led him to the unraveling of many of nature's mysteries and as a habit of thought made Morgan a sympathetic person, shunning nationalism, fanaticism, and prejudice of any kind. He was successful as a scientist, as a leader, and, even though Morgan would have scoffed, as a teacher. That he did it all with grace, modesty, enthusiasm, and unfailing humor and enjoyment is a tribute to his eccentric approach to science and to life.

Chronology

September 25, 1866   Morgan born in the family home, Hope-mont, Lexington.

Spring 1886   Bachelor of science degree, State College of Kentucky.

Fall 1886   Entered graduate program, Johns Hopkins University.

April 1888   First publication: "Experiments with Chitin Solvents," *Studies from the Biological Laboratory of the Johns Hopkins University.*

1888   Master of science degree, awarded in absentia, State College of Kentucky.

Spring 1890   Ph.D. degree, Johns Hopkins University.

1890–1891   Held Bruce Fellowship at Johns Hopkins University, during which time he traveled and made first visit to the Naples zoological station.

Fall 1891   Joined biology faculty, Bryn Mawr College.

1894–1895   Year's leave from Bryn Mawr spent at the Naples zoological station, working with Driesch and others.

1897   First book: *The Development of the Frog's Egg: An Introduction to Experimental Embryology* (New York: Macmillan).

1901   *Regeneration,* Columbia University Biological Series, vol. 7 (New York: Macmillan).

1903   *Evolution and Adaptation* (New York: Macmillan).

Fall 1904   Professor of experimental zoology, Columbia University.

1907   *Experimental Zoology* (New York: Macmillan).

1908   *Drosophila* entered the Morgan laboratory through the work of a graduate student, Fernandus Payne, and others. Morgan began efforts to induce mutation.

1909   President, American Society of Naturalists.

1910–1912   President, Society for Experimental Biology and Medicine.

July 22, 1910   First *Drosophila* paper: "Sex Limited Inheritance in Drosophila," *Science* 32:120–22.

September 10, 1911   "Random Segregation versus Coupling in Mendelian Inheritance," *Science* 34:384.

1913   *Heredity and Sex* (New York: Columbia University Press).

1915   *The Mechanism of Mendelian Heredity*, with A. H. Sturtevant, H. J. Muller, and C. B. Bridges (New York: Henry Holt).

1916   *A Critique of the Theory of Evolution* (Princeton, N.J.: Princeton University Press).

1916   *Sex-linked Inheritance in Drosophila*, with C. B. Bridges (Washington, D.C.: Carnegie Institution).

1916   Bridges published work on nondisjunction, providing additional proof of the chromosome theory of heredity and to the balance theory of sex determination in *Drosophila*.

1919   *The Genetic and the Operative Evidence Relating to Secondary Sexual Characters* (Washington, D.C.: Carnegie Institution).

1919   *The Physical Basis of Heredity*, Monographs on Experimental Biology (Philadelphia: J. B. Lippincott).

1920–1921   Visiting professor at Stanford University.

1922   *Some Possible Bearings of Genetics on Pathology* (Lancaster, Pa.: New Era Printing Co.).

1923   *Mechanism* revised and expanded.

1923   *The Third-Chromosome Group of Mutant Characters of Drosophila melanogaster*, with C. B. Bridges. Contributions to the Genetics of Drosophila melanogaster, no. 327 (Washington, D.C.: Carnegie Institution).

1923   *Laboratory Directions for an Elementary Course in Genetics*, with H. J. Muller, A. H. Sturtevant, and C. B. Bridges (New York: Henry Holt).

1924   *Human Inheritance* (Pittsburgh: University of Pittsburgh School of Medicine).

1925   *Evolution and Genetics* (Princeton, N.J.: Princeton University Press).

1925   *The Genetics of Drosophila*, with C. B. Bridges and A. H. Sturtevant (*Bibliographia Genetica* 2:1–262). This special journal issue, devoted to the Morgan-Bridges-Sturtevant studies, is sometimes called the Bible of geneticists and is in many respects a compilation of the important *Drosophila* work in earlier Carnegie Institution publications and else-where.

1926   *The Theory of the Gene* (New Haven, Conn.: Yale University Press).

1926   *Genetics and the Physiology of Development*, Fifth William Thompson Sedgwick Memorial Lecture (Woods Hole, Mass.: Marine Biological Laboratory).

1927   *Experimental Embryology* (New York: Columbia University Press).

1927–1928   President, National Academy of Sciences.

1928   Named head of the Division of Biology, California Institute of Technology.

1929   President, American Association for the Advancement of Science.

1929   *What Is Darwinism?* (New York: W. W. Norton).

1932   *The Scientific Basis of Evolution* (New York: W. W. Norton).

1932   President, Sixth International Congress of Genetics.

1933   Awarded the Nobel Prize in Physiology or Medicine.

June 4, 1934   Presented the Nobel Prize speech in Stockholm.

1934   *Embryology and Genetics* (New York: Columbia University Press).

1939   Awarded Copley Medal of the Royal Society.

1942   Retired, California Institute of Technology.

December 4, 1945   Died in Pasadena, where his body was cremated.

Notes

Chapter 2

1. The custom of these popular lectures continued far into the twentieth century. About 1916 Morgan is said to have illustrated a talk by projecting *Drosophila* on the screen. At the moment before showing, the etherized fruit flies were inserted into a double slide with an air space. The slides were then sealed. As the flies awakened the slide was projected, and huge, live *Drosophila* with their red eyes showing crawled briskly back and forth across the screen. The effect was startling and the audience loved it. The next slide was shown without delay for the flies were killed by the projector's heat within ten seconds after their performance.

Chapter 3

1. The Hopkins medical college admitted women, however, because a group of wealthy and influential women gave some $300,000 on the condition that women be admitted on an equal basis to men. But Hopkins did not generally admit women to its graduate programs until after 1907, giving it the dubious distinction of being the last of the great graduate institutions of the country to do so.

Chapter 4

1. The question that was seldom asked was whether they applied to man. For example, when Dr. Karl Landsteiner discovered in 1900 that the blood of his colleagues in Prague did not mix (because they had different ABO blood groups), it did not occur to him or to anyone else for eight years to wonder if the

differences in blood type obeyed Mendel's rules. However, when Archibald Garrod in 1902 described individual differences in human biochemistry, he appreciated (thanks to prompting by Bateson) that the biochemical individuality he was studying—a rare form of arthritis called alcaptonuria—was an example of Mendelian recessive inheritance. Without saying so explicitly, Garrod implied that genes made enzymes, which aided the metabolic process if they were normal, or halted it if they were abnormal. In the case of alcaptonuria, the normal complete destruction of homogentisic acid was blocked, thus leading to its excretion in the urine which "darkens with alkalis and on exposure to air." The clear statement that genes worked by producing enzymes and the proof that "what Garrod had shown for a few genes and a few chemical reactions in man, was true for many genes and many reactions" in all living things had to wait until the early 1940s when Beadle (Morgan's postdoctoral student and eventual successor at Caltech) and Tatum proved it by work on the red bread-mold *Neurospora* (work which began through studies on the chemistry and eye-color of *Drosophila*). Every bit as remarkable as Garrod's wonderful paper itself is the fact that it appears to have been ignored by researchers in the field, including both Beadle and Morgan, who often speculated about how genes worked. But in the 1920s and 1930s students heard about Garrod from teachers such as E. B. Ford, Larry Snyder, Charles Cotterman, and Sewall Wright. Perhaps the same was true in other classrooms. Not in Morgan's, however. Even in his later books, Morgan erroneously included eye color as part of his scanty list of established inherited conditions in humans that obeyed Mendel's laws, but continued to exclude alcaptonuria and albinism.

Our claim above that Garrod's work has been "utterly ignored" is simply our mistake that we prefer to allow to stand in order that the full force of Dr. Sewall Wright's comment to us should be appreciated: "The statement . . . is much too strong. Bateson played up Garrod and the enzyme theory of gene action strongly in 'Mendel's Principles of Heredity' which I read in 1913. It was probably read by all geneticists of that period. I took this theory for granted in my thesis (1916) in interpreting the peculiar gene interactions which I described in the guinea pig. It seemed to me (and others) that it was almost a logical necessity that genes operate by determining the specificity of enzymes. I did not feel that a general discussion of the theory was necessary in my thesis and thus did not refer to Garrod but merely referred to the papers that interpreted the interactions of the color factors of mammals in this way (Cuénot, Durham, Onslow, Gartner and others). Miss Durham probably made the first attempt to test the enzymatic properties of skin extracts from different color genotypes (of the mouse). Onslow made the most extensive early experiments of this sort. I organized a course in Physiological Genetics largely about this hypothesis at the University of Chicago in 1929. I devoted several lectures to Garrod's studies and many others to similar ones. Several of my students made skin extracts of diverse color genotypes of the guinea pig from this standpoint. The first to reach publication was W. L. Russell's thesis, 1939. In 1929 I questioned R. A. Fisher's theory that the prevailing dominance of wild type genes over deleterious mutations (of which even the heterozygotes were rare in nature) was due to natural selection of specific modifiers of these heterozygotes (selection pressure of the order 10^{-6} as we both agreed). I maintain that the phenomenon could be explained more plausibly as a byproduct of selection of modifiers of the abundant wild type homozygotes, reflecting the dosage-response curve of an enzymatic reaction limited by the rate of production of its substrate. Discussion got nowhere because of persistent misrepresentation of my hypothesis by Fisher and his associates. The great achievement of Beadle & Tatum in the 1940's was to correlate systematically the steps in the basic metabolic reactions with genes, and in some cases demonstrate enzyme differences."

Chapter 5

1. BRIDGES: Oh I've got a lovely gynandromorph—male on one side and entirely female on the other.

MORGAN (*from the neighboring desk*): Is that your conception of loveliness?

2. Readers interested in more of the details of this debate should see the authors' article on the topic to be published in *Journal of Heredity* (1976).

3. Dr. Charles Cotterman makes an interesting point about

the expression "Mendelizing together": "Morgan's use of the expression . . . is apt to cause puzzlement if we recall that Mendel by chance never discovered an instance in which two nonallelic genes failed to show random assortment. Thus characters which 'Mendelize together' are actually those that do not Mendelize."

4. Darwin's term *sex-limited*, referring to a mode of inheritance in which one of a pair of alternate traits is entirely restricted to one sex, was in early years used somewhat loosely by Morgan. But by 1913 he left the term as defined, in which sense it is still used, and introduced the term *sex-linked* to refer to "a character following the known distribution of the X chromosomes," such as white eye in *Drosophila* or color-blindness in man. Bridges called this criss-cross, which more vividly describes it.

5. Within five to ten years most geneticists had accepted the Morgan group's work, although as late as 1925 *Science* published an anti-Morgan editorial. The dogmatic refusal of communist scientists to accept Mendel-Morgan genetics, and the tragic consequences of their refusal, are detailed in Zhores Medvedev's *The Rise and Fall of T. D. Lysenko* (New York: Columbia University Press, 1969).

But it also needs to be mentioned in passing that this great book of Morgan's, like many great books, perpetuated its drawbacks. Flies from Columbia stocks were sent all over the world, to anyone who asked for them, and genetics became almost synonymous with *Drosophila*. It was sometimes said, if you were interested in rabbit genetics, you should study rabbits, but if you were interested in genetics, you should study *Drosophila*. And if you were interested in human genetics, you should forget it. When the American Society of Human Genetics was formed in 1948, most geneticists, including Muller and Curt Stern, were opposed to the idea—as Morgan would have been also. As a consequence of the success with *Drosophila*, attention was diverted from biochemical work and especially medical genetics and not redirected there until World War II. Although much of the history of genetics has been a story of stops, starts, and lengthy lapses, beginning with the submersion of Mendel's theories in 1865, the slowness of the science to take up medical questions was especially sad since much misery could have been avoided.

Chapter 6

1. Sturtevant lists Morgan's postdoctoral foreign students as Mohr, H. Nachtsheim, C. Stern, G. Bonnier, T. Komai, E. Gabritschevsky, T. Olbrycht, A. Zulueta, Y. Imai, T. Dobshansky. Others he includes in the Fly Room are F. N. Duncan, E. Cattell, E. Altenburg, J. S. Dexter, A. Weinstein, J. W. Gowen, D. E. Lancefield, and E. G. Anderson. Sturtevant also mentions Mrs. Morgan, although others insist she was never a genuine part of the Fly Room group. Carlson also lists among the *Drosophila* workers John Specter, Shelley R. Safir, Roscoe R. Hyde, and L. S. Quackenbush; workers between 1916 and 1920, he says, include also Harold Henry Plough, Charles Robert Plunkett, and Alfred Francis Huettner.

2. A letter Morgan wrote in 1923 to a friend and former student about the balance theory shows the give and take of the laboratory—and also Morgan's usual tendency not to commit himself the one step beyond the facts to theory: "I entirely agree with the criticism of the inadequacy of the theory of balance, of which Bridges is so inordinately fond. It is, and must remain a fiction so long as we cannot attach any objective values to the unit or units involved. It may be quite true and probably is in a way, but it tells us nothing more than the facts do themselves. This I have urged on Bridges from the beginning without any effect."

Chapter 8

1. For example, had Morgan read Schrödinger's proposal that the "gene or perhaps the whole chromosome fibre might be an aperiodic crystal, the structure of which forms a code script with almost unlimited possible molecular combinations," he would have scoffed at this unsubstantiated speculation, later to prove true.

Chapter 9

1. Linkage was first seen by Bateson and Saunders, although they failed to discover what they saw. W. S. Sutton, however, had predicted linkage and correctly interpreted Bateson and Saunders's results.

A Note on Sources

ONE CONTINUING measure of Thomas Hunt Morgan's forcefulness and likability is the generosity shown to the authors by his family, students, and former colleagues. All seemed happy to spend much time talking about Morgan, to our considerable advantage as biographers. Because these conversations, like those of specialists in the several fields in which Morgan worked, have been influential in the shaping of our conceptions of the man, we have included references to them in this section of our book. We must mention here also the following persons who read and criticized in detail one and sometimes two drafts of the manuscript: Dr. George and Mrs. Muriel Beadle, Dr. James Bonner, Dr. Charles Cotterman, Dr. Victor McKusick, and Dr. Sewall Wright. Contributions by Drs. Beadle, Cotterman, and Wright are so identified in the text. For help on taxonomy we are grateful to Dr. Mary Wharton and again to Dr. Cotterman. Mrs. Elizabeth Howard, Dr. Jacqueline Bull, and Joel Beane were very helpful with source materials.

Collected reprints and editions of Morgan's published work are available in their entirety at the Marine Biological Laboratory library at Woods Hole, Massachusetts. A fairly complete set is available at the University of Kentucky, the American Philosophical Society, and Caltech.

Lexington

The Hunt-Morgan Family Papers, 1784–1949, are housed in the Special Collections of the Margaret I. King Library, University of Kentucky, and include a small number of letters from Thomas Hunt Morgan and from Lilian Morgan to THM's mother in Lexington. The most interesting correspondence, however, is from THM's father Charlton to his wife and to their other son, Charlton.

The Lexington Public Library, and to a smaller extent Margaret I. King Library at the University of Kentucky, have microfilmed copies of Lexington newspapers contemporary with THM and his parents. Of greatest value to us were the *Lexington Transcript*, particularly for accounts of the continued celebration of John Hunt Morgan, and the *Lexington Morning Herald* and the *Lexington Herald* for stories after the turn of the century such as family obituaries and the celebration plans for THM's seventieth birthday.

A family tree more detailed than the one we show (back endpaper) has been prepared by the staff at the Hunt-Morgan House in Lexington. A copy of another, compiled by members of the Morgan family, is on file at the Thomas Hunt Morgan Institute of Genetics in Lexington. The most compact source of family history is James A. Ramage, "The Hunts and the Morgans: A Study of a Prominent Kentucky Family" (Ph.D. diss., University of Kentucky, 1972). Mr. Richard DeCamp was helpful in determining the accuracy of local stories.

There is no shortage of books dealing with John Hunt Morgan, THM's famous uncle. He has been studied strategically, turned into fiction and ballad, and written about from every point of view. Most helpful to us in our understanding of the general, his brother Charlton, and that earlier period of history were: J. Winston Coleman, *Lexington during the Civil War* (Lexington: Commercial Printing Co., 1938); Burton Milward, *The Hunt-Morgan House*, with architectural notes and drawings by Clay Lancaster (Lexington: The Foundation for the Preservation of Historic Lexington and Fayette County, 1955); and James W. Milgram, "The Rebel Raider, John H. Morgan, and His Men as Prisoners," *Confederate Philatelist* (July-August 1973): 91–104.

The best source of information on THM's boyhood days has been his family, whose encouragement and contributions to this work have been extremely helpful. THM's eldest child and only son, Howard Morgan, prepared a 7-page typed booklet *The Thomas Hunt Morgan Story* for the dedication of the Thomas Hunt Morgan Intermediate School in Seattle, Washington, on December 18, 1953. We also have been privileged to have had a long and friendly correspondence with Mr. Morgan and with THM's three daughters, Mrs. Edith Whitaker, Mrs. Lilian Scherp, and Dr. Isabel Mountain, all of whom read the manuscript in draft. Also helpful was Mr. Calvin Applegate, a cousin of Morgan. Copies of the typed booklet and all correspondence are on file at the Thomas Hunt Morgan Institute of Genetics, Lexington.

Morgan's boyhood and youth are included in two biographical accounts of his life: Herbert Parkes Riley, "Thomas Hunt Morgan," *Transactions of the Kentucky Academy of Science* 53 (June 1974):1–8; and Wendell H. Stephenson, "Thomas Hunt Morgan: Kentucky's Gift to Biological Science," *The Filson Club Historical Quarterly* 20 (1946):97–106.

The standard history of the University of Kentucky, including its State College of Kentucky days when THM was enrolled, is James F. Hopkins, *The University of Kentucky: Origins and Early Years* (Lexington: University of Kentucky Press, 1951). Also helpful were the annual registers of State College and the proceedings of the semicentennial celebration held in 1916. We are indebted to Dr. Dewey Steele, who knew both THM and an earlier Lexington, for a clearer understanding of THM's relationship to Crandall and the Peters—and also for a somewhat divergent viewpoint that Morgan may have been ashamed of his uncle, who after all was on the losing side and through his exploits forced his immediate family and even THM's sister Nellie to defend him all their lives.

For a clearer glimpse of the scientific climate of Lexington when THM was a boy and student, we referred to the Robert Peter Collection, part of the Howard Evans Collection in the Special Collections of Margaret I. King Library, University of Kentucky, and we sought the advice of Dr. John Ellis, Department of History, LeHigh University.

Johns Hopkins

Daniel C. Gilman, Hopkins's first president, described the university's beginnings in "The Launching of a University," in *Portraits of the American University 1890–1910*, compiled by James C. Stone and Donald P. NeDevi (San Francisco: Jossey Bass, 1971). Material concerning the training of biology students comes from Hugh Hawkins, *Pioneer: A History of The Johns Hopkins University, 1874–1889* (Ithaca, N.Y.: Cornell University Press, 1960).

Very little correspondence is available from THM during this period, and his attitudes toward work are taken largely from his published writings in the years immediately after graduation.

For some of the perspectives on biology at the time, both at Hopkins and at the Marine Biological Laboratory at Woods Hole, we followed Garland Edward Allen, "The Early Development of Genetic and Evolutionary Theory in America: A Study of Thomas Hunt Morgan" (Ph.D. diss., Harvard University, 1966). Dr. Allen's forthcoming book on THM is noted under Caltech.

Woods Hole

THM's personal comments come from his own letters and from his family's recollections. Discussion of the Marine Biological Laboratory comes from: Frank R. Lillie, *The Woods Hole Marine Biological Laboratory* (Chicago: University of Chicago, 1944); and Edwin Grant Conklin, "Early Days at Woods Hole," *American Scientist* 56 (Summer 1968):112–29. Conklin was a student at Hopkins, and this reminiscence also gives an interesting view of Morgan's teachers. For a more recent description of the laboratory, we are grateful to MBL librarian Jane Fessenden.

Bryn Mawr

THM himself, and later Lilian, provide the most informative material about the Bryn Mawr years in their letters to THM's mother and sister. The Bryn Mawr catalogs indicated teaching schedules. Also helpful was an article by Alice Boring (class of 1904), "Thomas Hunt Morgan," in the *Bryn Mawr Alumnae Bulletin* (February 1946), and correspondence from Mary Gardiner, formerly of the Bryn Mawr faculty.

For a better understanding of the embryological work of this period and also of Morgan's contribution to it, we consulted Dr. Jane Oppenheimer, both in conversation and in her writing, particularly: "Embryological Concepts in the Twentieth Century," in *Survey of Biological Progress 3*, ed. Bentley Glass (New York: Academic Press, 1957), pp. 1–37; "Embryology and Evolution: Nineteenth Century Hopes and Twentieth Century Realities," *Quarterly Review of Biology* 34 (1959):271ff; and *Essays in the History of Embryology and Biology* (Cambridge,

Mass.: M. I. T. Press, 1967). Dr. Dorothea Rudnick was also very helpful. Of much value too were Joseph Needham's *History of Embryology* (Cambridge, Eng.: At the University Press, 1940) and C. M. Child's *Patterns and Problems of Development* (Chicago: University of Chicago Press, 1941). See *American Naturalist* 35 (1901):973 for THM's flirtation with vitalism. Dr. Richard Goss—and his *Principles of Regeneration* (New York: Academic Press, 1969)—was of great value.

Columbia

Edmund B. Wilson announced the establishment of the chair of experimental zoology that brought Morgan to Columbia—and carefully spelled out the university's expectations both of Morgan and of experimental zoology—in "The Department of Zoology," *Columbia University Quarterly* 6 (March 1904):146–57. Also informative were: Henry E. Crampton, *The Department of Zoology of Columbia University, 1892–1942* (New York: Morningside Heights, 1942); Leslie C. Dunn, "Genetics at Columbia," *Columbia University Quarterly* 30 (December 1938):258–66; *A History of Columbia University, 1754–1904* (New York: Columbia University Press, 1904); and *Zoology: A Report on Research by the Department of Zoology for the Years 1919–1924* (New York: Columbia University Press, 1925).

Genetics: Introduction and History

For the general reader who would like to know more about genetics in a painless way, the following are excellent introductions: George and Muriel Beadle, *The Language of Life: An Introduction to the Science of Genetics* (Garden City, N.Y.: Doubleday, 1966); Aaron E. Klein, *Threads of Life* (Garden City, N.Y.: Natural History Press, 1970); Adrian M. Srb, Ray D. Owen, and Robert S. Edgar, *General Genetics*, 2d ed. (San Francisco: W. H. Freeman and Co., 1956). A basic genetics book giving more emphasis on medicine is Victor A. McKusick, *Human Genetics* (Englewood Cliffs, N.J.: Prentice-Hall, 1969).

Other, usually more technical, histories and surveys of genetics used in the writing of the book include: F. A. E. Crew, *Foundations of Genetics* (Oxford: Pergamon Press, 1966); L. C. Dunn, *A Short History of Genetics* (New York: McGraw-Hill, 1965); Robert Olby, *The Path to the Double Helix* (New York: Macmillan, 1974); and A. H. Sturtevant, *A History of Genetics* (New York: Harper and Row, 1965). In trying to focus almost completely on Morgan himself within our space limitations, we have given far too little attention to the controversy that continued among Morgan, Bateson, Castle, and others over the gene. This is well covered in Elof Axel Carlson, *The Gene: A Critical History* (Oxford: Pergamon Press, 1966). We enjoyed, as well as relied upon, *The Mendel Newsletter: Archival Resources for the History of Genetics & Allied Sciences* issued by the Library of the American Philosophical Society.

The Fly Room

The traditional description of the give and take of the Fly Room comes here, as in most briefer mentions of Morgan elsewhere, from Alfred H. Sturtevant's lengthy obituary *Thomas Hunt Morgan, 1866–1945: A Biographical Memoir,* (New York: published for the National Academy of Sciences of the United States by Columbia University Press, 1959). A much different viewpoint, stressing the conflict of ideas and personality clashes, particularly between Morgan and Muller, is presented in Elof Axel Carlson's "The *Drosophila* Group: The Transition from the Mendelian Unit to the Individual Gene," *Journal of the History of Biology* 7 (Spring 1974):31–48. It should be accepted, Carlson concludes, that "Morgan's role in the *Drosophila* group has been romanticized and overinflated, and that Sturtevant's interpretation is severely biased by the favored position he held throughout the long association he had with Morgan" (p. 48).

In attempting ourselves to assess Morgan's contributions in the field of genetics, whether individually or as a leader of the team, we have relied largely on his written work—a considerable undertaking made easier by the complete bibliography given in the Sturtevant memoir mentioned above. We discovered only one omission: Morgan's 1907 presentation to the International Zoological Congress of "The Role of Irritability and Contractility as Dynamic Factors in Development and Regeneration" (Cambridge, Mass., 1910), and one error of inclusion, "The Genetics Controversy" (1940), which was not written by Morgan. A few entries give the wrong date. Also of interest, and unnoted by Sturtevant, is the fact that many of the Morgan articles originally printed in *Archiv für Entwickelungsmechanik der Organismen* are reprinted and more easily available as Bryn Mawr Monographs.

We had many conversations with scientists who knew Morgan himself or historical genetics or both. These persons, to whom we owe much gratitude, include: Dr. Leland Brown, Dr. George Beadle, Dr. Charles Cotterman, Dr. James Crow, Dr. C. D. Darlington, Dr. E. B. Ford, Dr. Richard Goss, Dr. Philip Levine, Dr. Edward B. Lewis (THM Prof. of Genetics, Caltech), Dr. Joseph Needham, Dr. Robert Olby, Dr. Dorothea Rudnick, Dr. Laurence Snyder, Dr. Dewey Steele, Dr. Curt Stern, and Dr. Sewall Wright. Of particular help was Dr. Fernandus Payne, who has written of his own *Drosophila* work at Columbia in his delightful *Memories and Reflections* (Bloomington: Indiana University Press, 1974) and who has been generous with his time in conversations. Mrs. Phoebe Sturtevant, widow of Alfred Sturtevant, also kindly talked with us and wrote of her recollections of this period.

Other personal accounts of the Fly Room are contained in "The Reminiscences of Theodosius Dobzhansky" (1962) and "The Reminiscences of L. C. Dunn" (1961), both in the Oral History Collection of Columbia University; T. Komai, "T. H. Morgan's Times; a Japanese Scientist Reminisces," *Journal of Heredity* 58 (September 1967):247–50; and correspondence from Professor Dr. Konrad Lorenz, Österreichische Akademie der Wissenschaften, Institut für vergleichende Verhaltensforschung. Bernard Jaffe wrote an admiring chapter on THM, the subtitle of which was "American Science Comes of Age," in *Men of Science in America* (New York: Simon and Schuster, 1944). He had Morgan's help and approval and the chapter consequently gives a clear view of how Morgan viewed his achievements and those of his colleagues. It also contains a clearly written narrative of the various experiments in the *Drosophila* work. The Lutz book including the account of a white-eyed fly is *A Lot of Insects: Entomology in a Suburban Garden* (New York: G. P. Putnam's Sons, 1941).

Background material for this period of Morgan's work comes from Arthur Hughes, *A History of Cytology* (London: Abelard-Schuman, 1959); Joseph Needham, *Biochemistry and Morphogenesis* (Cambridge, Eng.: At the University Press, 1966); and several fine studies by Garland Allen, including "The Experimental Method in Biology: T. H. Morgan and the Theory of the Gene," *Synthese* 20 (1969):185ff; and "T. H. Morgan and the Emergence of a New American Biology," *Quarterly Review of Biology* 44 (1969):168–88. Also helpful were Morgan's obituaries of Wilson, Bateson, Nettie Stevens, Bridges, and other colleagues of this period.

Morgan's correspondence of the Columbia period is housed at the American Philosophical Library in Philadelphia, where many papers of William Bateson, Ross G. Harrison, and E. B. Wilson are also of interest. In addition to this collection, we were able to see the Morgan-Loeb correspondence at the National Institute of Medicine in Washington, D.C. (H. J. Muller's letters—over 30,000 of them—are at Indiana University in Bloomington.)

At the Morgans'

Lilian's letters to her mother-in-law from this period, especially from 1904 to about 1907, are in the Hunt-Morgan Papers at the University of Kentucky, and Howard K. Morgan's pamphlet on his father also deals with the period. But we are most indebted to Mr. Howard Morgan and to Mrs. Whitaker, Mrs. Scherp, and Dr. Mountain for their long and careful letters to us detailing this period of their parents' lives, for several days spent with Dr. Mountain, and for the help of all the Morgan children throughout the writing and revising of this book. Private papers still held by the family, as well as those already given to the American Philosophical Library, were extremely helpful.

Dr. Tove Mohr, a retired physician, was the wife of Dr. Otto Mohr, Morgan's first foreign postdoctoral student. They remained lifelong family friends. Dr. Tove Mohr's address to the Morgan centennial celebration did much to add a spirit of Morgan to that meeting, and her time spent with us at her home in Norway and her generous correspondence has, we hope, done the same for this book. We are also grateful to her, as well as to Dr. Mountain, for many of the informal photographs.

Discussion of Lilian Morgan's laboratory work comes largely from Bernard Jaffe's chapter on Morgan in *Men of Science in America*. Other correspondence contributing to this chapter includes that of Edward James Mathews, a cousin of Morgan and an occasional dinner guest during the New York years.

Caltech

Much of Morgan's professional correspondence during his California years is housed in the California Institute of Technology Archives in Pasadena. We were unable to examine this material, although we had access to the part donated to the American Philosophical Society Library in Philadelphia and saw other small portions through the kindness of Dr. Judith Goodstein, archivist at Caltech. For information on this period of Morgan's life, we have relied on his professional writings, on conversations and correspondence with his children and the Mohrs, on comments from Dr. James Bonner, and on the extreme generosity of Dr. Garland Allen and Dr. and Mrs. George Beadle. Dr. Allen's full-length biography of THM, scheduled to appear in 1977 from Princeton University Press, will pay detailed attention to Morgan's work and, in the Caltech years, to his role as administrator and organizer of the biology division. Dr. Allen freely gave us access to a draft of his chapter on Caltech and said "take what you need." We took almost exclusively dates and descriptions of Morgan's administrative responsibilities. Dr. George Beadle was one of Morgan's first and most successful postdoctoral fellows at Caltech; his understanding of Morgan's goals and procedures at the new school— and his marvelous accounts of life there with Morgan as head— give this chapter what soul it has. Mrs. Muriel Beadle, writer and student in many fields, contributed to this book as a voluntary editor and a critic; we gratefully give both her and her husband credit and appreciation here.

The anecdote of the brandy that begins our book and some of the other stories surrounding Morgan's trip to Stockholm come from "The Reminiscences of Warren Weaver" (1961) in the Oral History Collection of Columbia University. The letter from

Lilian Morgan about her husband's absorption in the salivary gland chromosomes is a copy of the original in the possession of Dr. Tove Mohr.

Nobel Foundation and W. Odelberg, *Nobel, The Man, and His Prizes* (Amsterdam: Elsevier, 1972) provides information on THM's nomination and award and also very useful insights into the bases for choosing recipients.

Conclusion

Because space was limited, it was possible to give but the most meager reference to the role of Sutton, Wilson, Muller, Bateson, and scores of other great geneticists, although it was necessary to measure their contributions so that Morgan's could receive a relative appraisal. In this we were assisted by the histories of genetics and cytology mentioned above and also by Victor A. McKusick's "Walter S. Sutton and the Physical Basis of Mendelism," *Bulletin of Medical History* 34 (1960):487–97; by H. J. Muller's introduction to the recent reprint of E. B. Wilson's *The Cell in Development and Heredity* (New York: Academic Press, Johnson Reprints, 1966); by Alice Levine Baxter's "Edmund Beecher Wilson and the Problem of Development" (Ph.D. diss., Yale University, 1974); by E. A. Carlson's edited collections of H. J. Muller's essays, *Man's Future Birthright: Essays on Science and Humanity* and *The Modern Concept of Nature: Essays on Theoretical Biology and Evolution* (both Albany: State University of New York Press, 1973), by J. B. S. Haldane's review of Muller's selected papers and the papers themselves, *Studies in Genetics* (Bloomington: Indiana University Press, 1964). We are indebted to Dr. A. G. Cock for access to material on William Bateson, to appear in Dr. Cock's forthcoming book. Dr. Sewall Wright and Dr. George Beadle have been most helpful in providing general comments and an overall appreciation of this period.

Index